Propellers & Impellers: Research, Design, Construction and Application

London, United Kingdom
27-28 March 2019

ISBN: 978-1-5108-9561-4

Printed from e-media with permission by:

Curran Associates, Inc.
57 Morehouse Lane
Red Hook, NY 12571

Some format issues inherent in the e-media version may also appear in this print version.

The Royal Institution of Naval Architects
8-9 Northumberland Street
London, WC2N 5DA
United Kingdom

Phone: 020 7235 4622
Fax: 020 7259 5912

publications@rina.org.uk

Additional copies of this publication are available from:

Curran Associates, Inc.
57 Morehouse Lane
Red Hook, NY 12571 USA
Phone: 845-758-0400
Fax: 845-758-2633
Email: curran@proceedings.com
Web: www.proceedings.com

TABLE OF CONTENTS

Propellers & Impellers: Research, Design, Construction and Application

27 - 28 March 2019, London, UK

PAPERS

Sponsored by:

NUMECA INTERNATIONAL

THE ROYAL INSTITUTION OF NAVAL ARCHITECTS
8-9 Northumberland Street, London, WC2N 5DA, Telephone: +44 (0)20 7235 4622

USAGE OF DIGITAL TOOLS FOR THE OPTIMUM PROPELLER DESIGN

N W H Bulten, Wärtsilä, The Netherlands

SUMMARY

The usage of more sophisticated design and analysis tools have become prevalent in all industries. In order to get the true value out of these numerical tools, the actual design requirements need to be defined properly and in general, this leads to the question of how to find the right balance between conflicting requirements. More feedback on the performance of a design can be obtained with usage of numerical simulations within the design evaluation phase and with digital data collection tools in the operational phase. With the implementation of CFD simulations in the design cycle, unforeseen deviations with model testing campaigns and actual full scale sea trials has reduced, and collection of actual vessel operational data, in combination with a system simulation approach generates further insight in the actual operation, which can be used to further fine-tune the design philosophy. Detailed evaluation of the results from full scale CFD simulations have revealed interesting phenomena related to the Reynolds scaling effects of ducted propellers.

1. INTRODUCTION

The requirements for propeller designs are becoming even more stringent due to changing market circumstances and legislation. In order to limit the emissions as much as possible, there is a clear focus on the propulsive efficiency whilst on the other hand, the comfort on board for crew and passengers, as well as the underwater noise levels, are nowadays considered as critical design criteria. Setting the target design criteria for a propeller is one of the aspects in the design cycle however, another important aspect is the method (or the range) of evaluation conducted for the design. In the old days this evaluation was limited to experimental model scale testing of performance and cavitation behaviour. Nowadays the developments in numerical flow simulations (CFD) have reached a level where they can be adopted within the evaluation and feedback cycle, such as those in [1-4]. Despite the developments in both numerical and experimental evaluation methods, the final truth is found in the vessel's sea trials, where the ship speed, noise and vibration levels are determined hence, implying that there are three different possible feedback loops in the evaluation of the actual propeller design, as shown in figure 1.

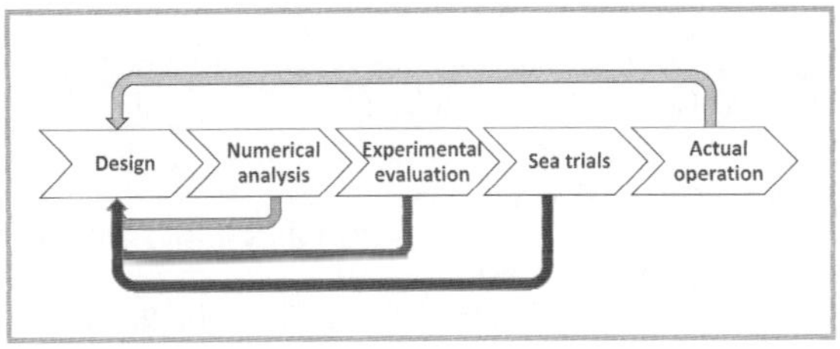

Figure 1: propeller design process and (possible) feedback-loops

The key-elements in feedback loops are the accuracy of the evaluation and the required time to obtain that feedback. It is obvious that changes in the design are costlier after sea trials, in comparison to changes during the drawing phase. Even when there are still discussions with regards to the accuracy of CFD simulations, value can be found in the shorter runtime of a range of various simulations, when compared to that of a model test campaign and furthermore, when considering the cost impact in comparison to full sea trials.

So far the optimum propeller design is based on meeting the targets during the sea trials. It should be acknowledged that the actual operation of the vessel will differ quite significantly from the defined sea trial condition, which is also related to whether the design is made for the owner or for the ship yard, since the owner will operate the vessel for many years and will regard the impact on fuel consumption as paramount. For such reasons, the feedback from vessel monitoring / data-collection systems can and will provide valuable insights into the actual performance of the vessel and propulsion system, which can be used to make a fourth additional feedback loop in the design cycle based on dynamic system simulations, thus aiding in meeting these sea trial requirements better.

The implementation of the digital tools, both CFD simulations and data analysis, have also triggered the discussion with regards to the impact of Reynolds scaling effects in the design [5]. The well-known ITTC'78 extrapolation method for full scale vessel performance [6] has been based on the model scale measurement data, a semi-empirical approach and limited full scale sea trial data. This methodology has worked quite well for about 40 years, though its limitations are becoming more visible for modern types of vessel designs and propulsion configurations.

2. CONFLICTING REQUIREMENTS

The fact that a vessel or propeller has conflicting design requirements is not unique in its own. With the current competitiveness across all industries, any design needs to address multiple requirements, targets and constrains.

In the propeller design cycle, there are two main requirements which are considered: maximum fuel efficiency and minimum impact of cavitation nuisance (either noise & vibrations or cavitation erosion).

2.1 PERFORMANCE AND COMFORT

Modern RoPax-vessels, as a typical example, have challenging propulsion unit designs. The versatile operational profile, with operation at various power ranges (single or twin engine power input operation) in combination with good manoeuvring capabilities and safety legislation, require a propeller design which can handle a large range of operating conditions.

The main characteristics of the propellers are the input power and the selected propeller diameter. In general, the propeller diameter is selected as large as possible to reduce the power density and in turn is a recognized method to improve efficiency. The propeller diameter will often be constrained by the target clearance between propeller tip and hull on the one hand and the vessel draft on the other. In case the clearance becomes too small, the impact of pressure pulses and noise will play a dominant role in the design.

Further limitations on efficiency arise from the impact of cavitation, which is one of the main design performance indicators of a propeller, next to the fuel efficiency. A sketch of a Pareto-front showing the propeller design trade-offs with different efficiency and cavitation behaviour is illustrated in figure 2. Due to the combination of regulations on both efficiency (EEDI) and comfort class (for example Lloyds PCAC – passenger and crew accommodation comfort [7]), the right balance between the two targets need to be found. With the utilization of CFD simulations, the design can be better balanced before the conventional model testing campaign is executed. The complexity of the CFD simulations should not be underestimated however, since it must contain the complete vessel geometry and shaft line, as well as the free surface and the cavitation modelling. A more detailed description of the numerical approach can be found in [8]. With this numerical approach, both the overall efficiency as well as the pressure pulse levels on the hull surface can be determined in the early stage.

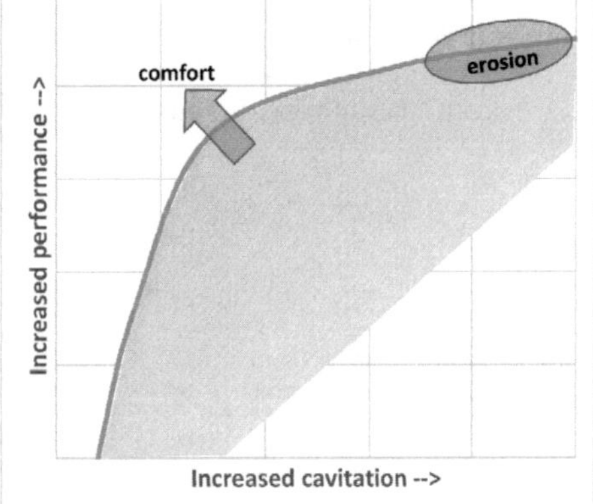

Figure 2: Pareto-front of conflicting design requirements

2.2 PERFORMANCE AND EROSION

Besides the challenges with RoPax-vessel propeller designs, there is another category of ships where the trade-off between efficiency and impact of cavitation is critical. This applies to vessels where the acceptable cavitation is close to the limit of undesirable levels of erosion however, by stretching the cavitation behaviour to its erosion limits, more efficiency can be achieved (such as reduced blade area and increased tip-loading). The conventional methods to evaluate the risk of cavitation erosion are based on observations of the cavitation cavity's shape development during a revolution or based on so-called paint-tests.

In 2015 the EU-funded Cafe project [9] started, where fundamental research has been carried out on the principles of cavitation erosion, amongst others in diesel injectors and ship propellers. The project combined both experimental and numerical methods to investigate different types of cavitation and its erosion, covering a full range from fundamental investigations to industry applications. One of the targets of the project had been to develop a numerical method to determine the risk of cavitation erosion on a marine propeller in behind-ship condition, based on RANS-CFD simulations. This cavitation erosion risk assessment is thus an extra step beyond the complex CFD simulations as mentioned before, when discussing the performance and pressure pulse levels. Initial results from the validation process have been presented by Melissaris *et al.* [10]

3. IMPACT OF REYNOLDS SCALING EFFECTS

The numerical flow simulations can be run both on model scale and on full (ship) scale to investigate the impact of the Reynolds scaling effects. Even though the validation of the numerical simulations is often based on model scale data only, good indications can be derived from the comparison of model scale and full scale results.

3.1 OPEN WATER PERFORMANCE

The analysis of Reynolds scaling effects for propellers in an open water set-up are split into two parts. First, the results of open propellers will be presented, followed by those of the ducted propellers. Where open propeller results confirm the existing opinion as used in the ITTC'78 method however, for the ducted propeller the outcome has been significantly different. Furthermore, research on full scale ducted propeller performance confirm the clear deviation from open propellers [11 - 13].

3.1 (a) Open propellers

In the search for the optimum propeller design, the performance has been identified as a key-parameter. The performance in uniform inflow (open water set-up) is determined as part of the overall performance evaluation process of the propeller although the actual interest is in the performance in the self-propulsion test, where the propeller is working in behind ship condition.

RANS-CFD results of the propeller open water performance calculated with SimCenter Star-ccm+ have been presented in detail [5, 14] With figure 3 showing the comparison of the open water performance based on both model scale CFD simulations and model scale measurements from various basins. It has been concluded that the prediction of open water performance at model scale can be done quite accurate, based on this population of different propeller designs. At the design point the spread is within 1%, which is of a similar magnitude as that of the spread over the model test experiments.

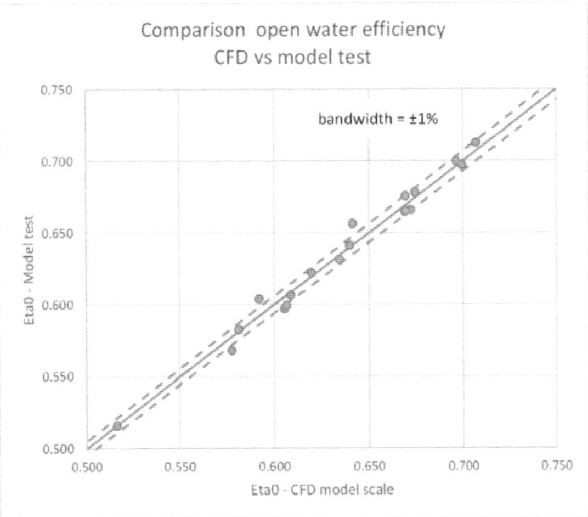

Figure 3: comparison of calculated and measured open water efficiency of open propellers

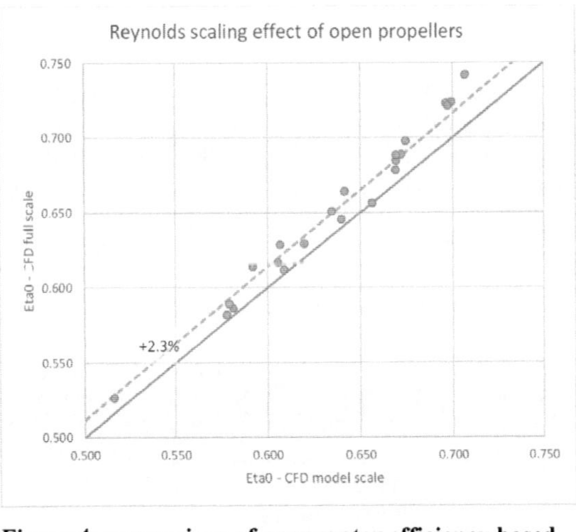

Figure 4: comparison of open water efficiency based on model scale and full scale CFD simulations

It should be acknowledged that the presented CFD simulations are based on a laminar flow regime, and the same propellers have also been analysed in a fully turbulent flow regime on model scale. The results of the full turbulent flow clearly indicate a reduction of open water efficiency. This reduction of open water efficiency is attributed to the difference in surface friction value between laminar and turbulent flow at model scale Reynolds numbers. At full scale the flow is expected to be fully turbulent, which results in a reduced friction coefficient, when compared with the model scale turbulent friction value, leading to increased efficiency. Figure 4 shows the comparison between the CFD results from model scale laminar flow simulations against those of full scale fully turbulent simulations. The average increase in efficiency from model scale to full scale is 2.3% using turbulent flow for this population of propellers.

3.1 (b) Ducted propellers

The open water performance of ducted propellers has been studied in detail as well. The CFD validation simulations have indicated that good agreement with model scale tests can be obtained when the turbulent flow regime is selected. The open water curves of a ducted propeller are shown in figure 5, based on model scale measurements, model scale CFD simulations and full scale CFD simulations.

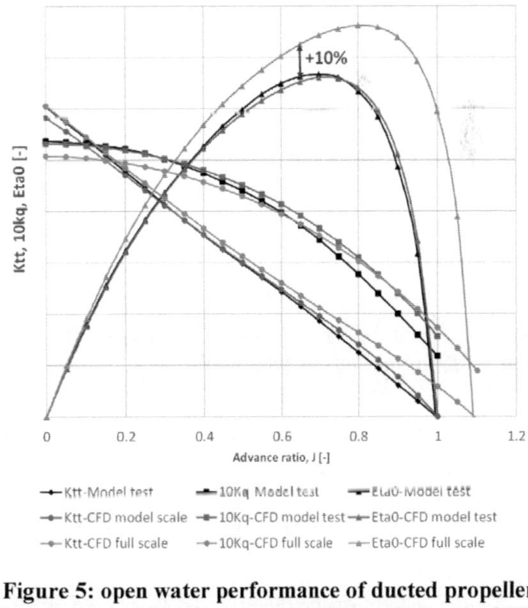

Figure 5: open water performance of ducted propeller based on model scale measurements, model scale CFD simulations and full scale CFD simulations

At full scale the surface friction factor reduces, which leads to improved performance, similar to those of an open propeller. In addition to this, the flow rate through the nozzle increases, which leads to a shift in the operating point of the propeller. The overall impact of the Reynolds scaling effect is a performance increase within10% for a ducted propeller. Another way to evaluate the differences in performance between model scale and full scale is based on the axial pump theory as presented in [2]. Reynolds scaling effects of axial pumps can be in the range of 7-8% [15], which explain the significant differences in performance, and confirm the 10% Reynolds scaling effect of ducted propellers in an open water set-up.

3.2 IN BEHIND-SHIP PERFORMANCE

The Reynolds scaling effect for ducted propellers in an open water set-up are claimed to be about 10% in free sailing condition. This statement deviates significantly from the ITTC'78 perception and it should be acknowledged that the actual open water performance is of limited interest, since the ducted propeller should perform well in behind-ship condition.

Based on model scale tests and the standard ITTC'78 scaling procedure, a prediction has been made for a tanker with a ducted propeller at the design power of 2870 kW, and vessel speed of 12.85 knots, shown in figure 6.

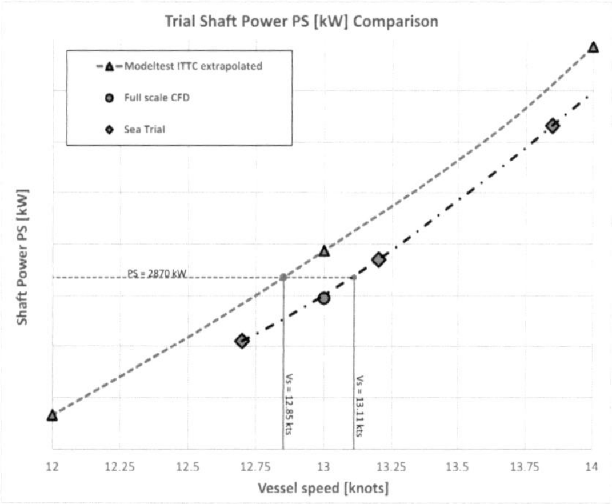

Figure 6: speed-power results from model test, full scale CFD simulations and sea trials

Moreover, a full scale CFD simulation of the complete vessel, including nozzle, propeller, brackets and rudder, has also been conducted for the design speed of 13 knots. The required power to reach self-propulsion is shown in figure 6, with an expected power level well below the available 2870 kW and thus from this CFD analysis, it can be concluded that the target design speed would be feasible for the current configuration.

Even though the ship speed prediction from model tests was below the target ship speed of 13 knots, it was decided to build the vessel according that design. The sea trials have confirmed that a ship speed above 13 knots could be achieved (13.11 knots) for the target power level. The sea trial curve as shown in figure 6 shows a very close agreement with the calculated point from the prior executed full scale CFD simulations.

The significant difference between prediction based on model scale measurements, the ITTC'78 extrapolation method, and the sea trial is another indicator that there is a large Reynolds scaling effect for ducted propellers.

The impact of this large scaling effect is not limited to the difference in observed versus expected vessel speed, with its related fuel savings, but also the underestimation of performance of ducted propellers in the range of 13-15 knots has an impact on the selection of the overall lay-out of the propulsion system in the early design phase. Selection of the conventional solution with an open propeller is thus a logical choice, as long as the performance evaluation is based on conventional model testing. Based on available feedback from both full scale CFD simulations and sea trials, the choice for open or ducted propellers for a certain application could be reconsidered. Once more vessels with ducted propellers are put into service, evidence of better performance on full scale will develop over time and furthermore confirm the shortcomings of the current model scale extrapolation method for ducted propellers as a logical result.

4. FROM SEA TRIAL TO ACTUAL OPERATION

As shown in figure 6, the design condition has been brought back to one single operating condition which is often not representative of the actual operation of the vessel for many years. In case of controllable pitch propellers, there can be a degree of freedom in the selection of operating pitch and RPM where desired thrust can be achieved with different combinations of these. Consequently, the required shaft power and fuel consumption depend on the drive-line layout and the engine characteristics.

Another aspect of the actual vessel operation is the occurrence of (unavoidable) dynamic events, which need to be considered.

4.1 VERSATILITY OF PROPELLER DESIGN

Controllable pitch propellers are often selected in case the vessel has a large variety in operational conditions or a high demand for manoeuvrability. Vessels with controllable pitch propellers are among twin screw RoPax/RoRo vessels, large anchor handling tugs and single screw fishing vessels. The variety in operation for RoPax/RoRo vessels can be found in summer and winter sailing mode for instance, where in winter season the time tables can be different, which allow for a slower sailing speed. In order to save fuel, both propellers can be run on a single engine, implying that the propeller has to be able to operate within a large power range at the same maximum RPM, whilst keeping performance and cavitation behaviour under control. For fishing vessels there is a clear difference between trawling at low speed and sailing to or from the fishing location at higher speeds. In trawling mode the required thrust clearly exceeds the hull resistance, which leads to an increased propeller loading condition. This effect is the same for tugs in pulling operation, where the majority of the thrust force is directly used to generate the towing force yet, in the free sailing condition propeller loading is much lower resulting in another operating point.

The choice for a controllable pitch propeller can be driven from a different requirement as well. Such as in case there is a need for a constant engine RPM operation, for example to drive a generator, the controllable pitch propeller can be selected to control the thrust force. From a fuel efficiency point of view, it is not recommended to always operate the propeller at constant RPM operation. Calculations of fuel savings when switching from constant RPM to variable RPM operation need to be made in order to make the proper evaluation, while considering the economical trade off against the additional costs of more complex, variable RPM generators.

4.2 DYNAMIC EVENTS

In real operation, the vessel will encounter many dynamic events, such as acceleration, manoeuvring and sailing in adverse weather conditions, where wind and waves start playing a role. Even the simplest dynamic event such as a vessel accelerating to its target speed can stir discussion around the actual impact of the dynamics in the system.

As long as the design is focussed on the steady-state operation at certain speed, the main parameter of interest is the hull resistance. In steady state it is assumed that the total thrust force is balanced with the total vessel resistance, yielding no acceleration of the vessel however, it should be acknowledged that in case the equilibrium between thrust and resistance is not achieved, there will be an acceleration of the vessel, based on Newton's law.

Various developments in the shipping industry have driven towards less installed power on board ships, for example due to improved hull resistance. The required thrust, and thus power, to accelerate the vessel to the target speed can become more critical in such situation, certainly when the vessel resistance is increased due to adverse weather conditions. A simple example case of a supply vessel with fixed-pitch propellers has been presented, where the issue of sustainable speed in adverse weather has been analysed [16]. A comparison between an azimuth thruster with open and ducted propellers has been made to indicate the impact of the different type of propulsion device on the overall vessel operability. Another example is the capability to accelerate a single screw vessel when equipped with either fixed pitch or controllable pitch propeller. In case of a fixed pitch propeller, the torque limiter of the engine will play an important role, whereas the controllable pitch propeller can be run up to maximum RPM before the pitch is increased. With this approach the full power from the engine can be made available much quicker.

4.3 SYSTEM SIMULATION APPROACH

In order to get more insight in the actual operating conditions of a propulsion system the system simulation approach is adopted. In such numerical model the performance characteristics of the various main components are modelled and the dynamic interaction between the different components is calculated.

4.3 (a) Main components of the system simulation model

A simple sketch of a system simulation model, as build in SimCenter Amesim, is shown in figure 7. In this model a number of main components are coupled with each other to determine the impact of the interaction.

The main components in this system are:

1. Hull resistance and inertia
2. Controllable Pitch Propeller performance response surface for thrust and torque as a function of inflow speed, propeller pitch setting and propeller RPM
3. Drive-line layout with the option to model a 2-speed gearbox and simplified 4-stroke engine
4. Operation/controls input channel

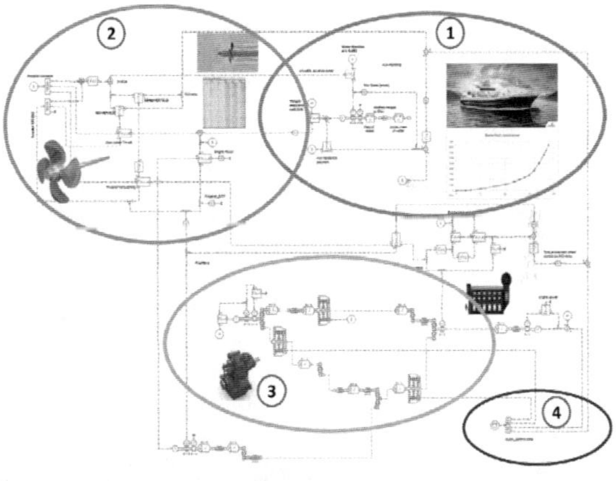

Figure 7: example of system simulation set-up for dynamic event simulations

The hull resistance will be balanced with the propeller thrust to reach the steady state sailing speed. The impact of the vessel inertia is taken into account, which gives a clear delay on the response of the vessel in case the thrust is changed.

The performance data of the propeller is based on a set of CFD simulations of the propeller in open water set-up for different pitch angles. The dimensionless thrust and torque data is translated into a mathematical response surface representation, where the propeller thrust can be determined as a function of ship speed and propeller pitch ratio, including the propeller-hull interaction factors for wake and thrust deduction. The propeller operating point will gradually change in the start-up phase until the target ship speed has been reached. The (translational) surge motion of the vessel is one part of the simulation model, where the longitudinal forces from the propeller are coupled to the resistance of the vessel. The second part of the model covers the rotational motions of the propeller, shaft-line and engine. The rotational motions are driven by the torques as provided by the engine and required by the propeller. The mean propeller torque is derived from the performance response surface, given the propeller RPM, pitch ratio and ship speed. The impact of the blade loading fluctuations due to the non-uniform wake is taken into account with a blade frequent torque fluctuation.

The complexity of a drive-line system can vary from a 2 stroke engine with a direct driven propeller to a modern system of a 4 stroke engine with gear-box, power taken in/out and battery systems. In the current system a simple drive-line modelling is applied which takes the inertia terms of the shaft-line, gearbox and engine into account. The current system simulation model has a simplified engine characteristic to monitor whether the torque limits are not exceeded.

4.3 (b) Validation of simulation model

For the validation of the dynamic system simulation model, a set of monitoring data from a single screw vessel has been made available. The validation simulations are driven by the time-dependent propeller operating condition, represented by propeller pitch-ratio P/D and propeller RPM. A typical example of a 10 minute time window is shown in figure 8, where both propeller pitch and RPM are varied. The signal for the propeller RPM shows several steps, which can be related to the handle position of the propulsion controls system. The signal for propeller pitch shows an irregular behaviour in the midsection of the simulation, which is related to the load limiter in the control system. The increase of propeller RPM of around 25% leads to a torque increase of more than 50% and almost doubling the total power.

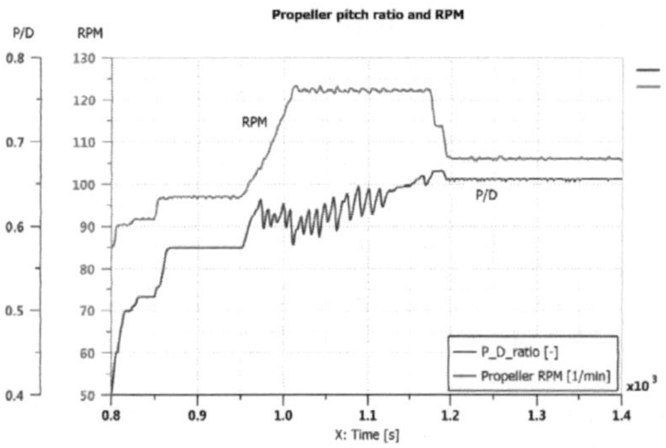

Figure 8: Time-dependent propeller operating condition (P/D & RPM)

The response of the system is calculated based on these two input signals and two main values which are used to validate the system simulation set-up are the required propeller shaft torque and the achieved ship speed.

The system response of the propeller shaft torque is shown in figure 9, based on the propeller operating conditions as discussed above. The steps in propeller RPM variation can be recognized in the response of the propeller torque and agreement between measured and calculated torque signals is reasonable over the whole time series. The blade-frequent torque fluctuations are visible in the diagram as a bandwidth around the torque signals.

The calculated ship speed is shown in figure 10, where the accelerated motion of the vessel can be observed. This acceleration is a result of the imbalance between thrust and resistance up to the point around 1200 seconds, as shown in figure 11. The increase in ship speed has an impact on the working point of the propeller, which can be observed in the time range from 850 to 950 seconds where propeller pitch and RPM remain constant, though the propeller torque steadily drops. Once the target ship speed has been reached at around 1200 seconds, the propeller RPM is reduced quite significantly, and the shaft torque and propeller thrust drop consequently. At this point the vessel resistance exceeds the delivered propeller thrust as shown in figure 11 and the equilibrium is restored when the ship speed is reduced to a speed of 12 knots. The large area in between the two force lines from 800 till 1200 seconds represent the additional required thrust to accelerate the vessel. The impact of the inertia of the vessel is a critical part of the dynamic system simulations, which can easily be overlooked in steady state analyses.

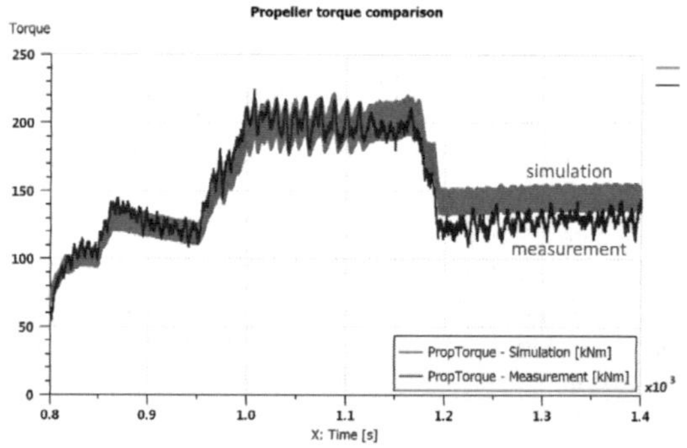

Figure 9: Comparison of measured and calculated propeller torque

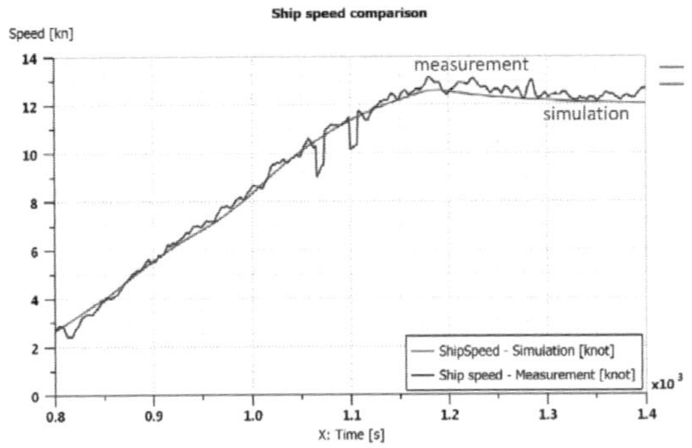

Figure 10: Comparison of measured and calculated ship speed

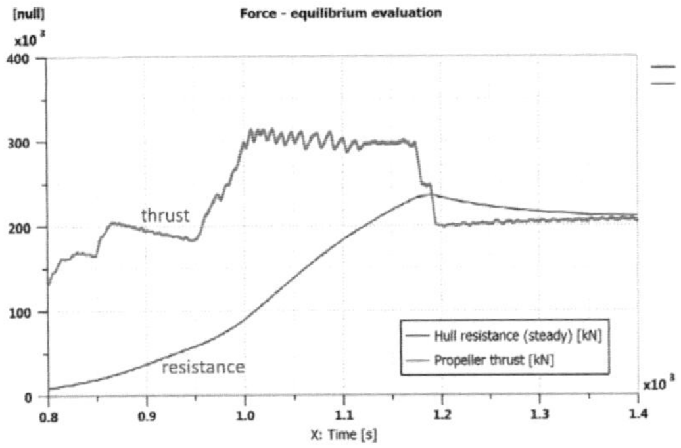

Figure 11: Propeller thrust - hull resistance balance

4.3 (c) Added value from system simulations

Once the dynamic simulation model has been validated to confirm the outcome, various operational scenarios can be analysed and compared, such as the handle settings for a controllable pitch propeller, which is denoted as combinator curve. This curve gives a set of propeller pitch-RPM combinations from full ahead to full astern handle position. The values for such curves are chosen to cover the complete vessel operational range. The required shaft torque for each handle position needs to be aligned with the available engine torque to avoid engine overloading. On the other hand, the optimal engine operating region is taken into account to obtain the best fuel consumption.

4.3 (d) Future developments

The currently presented model has several limitations which will be improved in a future version. Now, the hull resistance is based on the calm water hull resistance and the impact of adverse weather, wind and waves can be added in the part where hull resistance forces are being calculated. Fine-tuning of the model can also focus on the implementation of actual propeller-hull interaction factors in the propulsion condition. The current model is based on wake-fractions and thrust-deduction factors as derived from a limited number of steady-state conditions.

Another extension of the model will be related to the implementation of rudder forces and manoeuvring capabilities to be able to simulate other scenarios than straight sailing only. During the validation process of the current version, it has been observed that the impact of current can also play a role, certainly when reviewing the achieved vessel speed in comparison to the measured speed. Once the vessel starts to drift, the impact of a varying wakefield on performance might play a role as well and further integration and coupled simulations might be necessary to simulate the dynamics properly.

Further extensions are expected on the drive-line side as well, when more complexity is added. Typical developments focus on the transition from direct Diesel-mechanical drive-line layouts to systems with Diesel-electric and hybrid systems. The module where the gearbox is modelled can be extended with electric motors, driven either through auxiliary generator sets or batteries.

5. CONCLUSIONS

The demanding market situation in combination with more stringent regulations demand the usage of the best available tools in the design and evaluation process. In order to reach the goal of optimum design, it is of key importance to properly define the target design criteria. In general, the different criteria, such as efficiency, noise and vibrations, strength and durability have conflicting impacts when changing varying the propeller geometry. Finding the right compromise between the different design targets results in a multi-parameter optimization process.

It has to be acknowledged that both numerical and experimental design evaluations may have their shortcomings in the performance evaluation of a design variant, and it is therefore recommended to combine the use of all available methods. The feedback from each methodology, either potential flow-based methods, RANS CFD or model scale measurements should all be used to enhance the understanding and to improve the design further.

An example of the utilization of the feedback from sea trials confirms the significantly larger Reynolds scaling effects in ducted propellers, with acclaimed 10% performance increase from model scale measurements to full scale performance, found and explained in detail based on simple physical principles.

The current method to evaluate the performance of ship and propeller is still based on sea trials however, it should be acknowledged that this is not the best practice to identify the best propeller for actual use. Many ship operations have confirmed this by retrofits of propellers on existing fleet. The business cases for retrofits clearly benefit from the data monitoring of a sailing fleet in combination with the increased understanding of the actual operational profile.

6. ACKNOWLEDGEMENTS

The contributions to this work of both the Hydrodynamics and the System & Product Performance teams at Wärtsilä Propulsion are highly appreciated.

7. REFERENCES

1. PONKRATOV, D., ZEGOS, C., 'Validation of ship scale CFD self-propulsion simulation by the direct comparison with sea trials results', *Proceedings 4th International Symposium on Marine Propulsors*, Austin, TX, June 2015
2. BULTEN, N., 'With numerical simulations to more efficient ship designs', *Proceedings RINA Energy Efficient Ships conference*, London, UK, 2016
3. SAYDAM, Z., GOKCAY, S., INSEL, M., 'CFD based vortex generator design and full-scale testing for wake non-uniformity reduction', *International Journal of Naval Architecture and Ocean Engineering*. Vol 153. pp 282-296. 2018
4. JASAK, H., VUKCEVIC, V., GATIN. I., LALOVIC, I.,'CFD validation and grid sensitivity studies of full scale ship self propulsion', *International Journal of Naval Architecture and Ocean Engineering, Vol 11, 2019.*
5. BULTEN, N., STOLTENKAMP, P.,' Full scale CFD: the end of the Froude-Reynolds battle', *Proceedings 5th International Symposium on Marine Propulsors*, Espoo, Finland, June 2017
6. https://ittc.info/media/1593/75-02-03-014.pdf
7. https://www.lr.org/en/rules-and-regulations-for-the-classification-of-ships/
8 BULTEN, N., 'Full scale performance evaluation of ferries with CFD simulations', *proceedings FAST2017 conference,* Nantes, 2017
9. http://cafe-project.eu/
10. MELISSARIS, T, BULTEN, N, VAN TERWISGA, T, 'On Cavitation Aggressiveness and Cavitation Erosion on Marine Propellers using a URANS Method', *proceedings CAV2018 conference,* Baltimore, 2018.
11. ABDEL-MAKSOUD, M., HEINKE, H.-J.: 'Scale Effects on Ducted Propellers', *24th Symposium on Naval Hydrodynamics*, Fukuoka, Japan, July 2002
12. RIJPKEMA, D., VAZ, G., "Viscous Flow Computations on Propulsors: Verification, Validation and Scale Effects", *Proceedings of RINA-CFD2011*, London, UK, March 2011.
13. BHATTACHARYYA; A., KRASILNIKOV, V., STEEN, S., 'Scale effects on open water characteristics of a controllable pitch propeller working within different duct designs', *International Journal of Naval Architecture and Ocean Engineering*, Vol 112, pp 226-242, 2016
14. BULTEN. N.W.H., 'Transient blade load determination in behind ship based on CFD', *SNAME Propeller & Shafting conference,* Norfolk, 2015
15. PFLEIDERER, C.,'Kreiselpumpen', 5th ed. Springer Verlag, 1961
16. BULTEN, N. STOLTENKAMP, P., 'Improved sustainable speed due to thrusters with ducted propellers, *Proceedings of the ASME 2017 36th International Conference on Ocean, Offshore and Arctic Engineering OMAE2017*, Trondheim, Norway, 2017

8. AUTHORS BIOGRAPHY

Norbert Bulten holds the current position of Product Performance Manager at Wärtsilä – Business line Propulsion. He is responsible for hydrodynamics within Propulsion business. He has more than 20 years of experience in the field of CFD and marine propulsion systems, ranging from azimuth thrusters for DP-systems to high speed waterjets. In his current assignment he focusses on development and implementation of system simulations and advanced CFD-methodologies.

FULL SCALE PREDICTION OF HYDRODYNAMIC NOISE USING COMPUTATIONAL FLUID DYNAMICS

E. M. Fay, Cotty Fay Marine Design Inc. USA

SUMMARY

The issues of noise and vibration related to the propulsion system of vessels shines a bright light over the present day operation of sea going vessels. Some of the principal issues coming to the front are not new and are concerned with passenger and crew comfort which has been a touchstone for a decade or more. The most recent, more ominous issue is transmission of noise from transiting vessels and how this noise affects marine mammals.

The tools used to analyse and define the expected levels of vibration and noise in the design phase are becoming more and more robust. The paper describes the use of computational fluid dynamics to predict the noise and vibration generated by hydrodynamic flow over the hull and the propeller(s) of the vessel. The analysis is carried out using the program OpenFOAM comparing the operating propeller performance coefficients with the open water propeller coefficients. The paper also looks at the effects of cavitation, vessel trim and propeller loading on a 140 meter car ferry. The loading, noise and vibration data will be quantified and compared to full scale vessel data.

1. INTRODUCTION

The state of the art in predicting the noise levels within the ship for the comfort of crew and passengers is relatively well established. The prediction of the vessel's transmission of noise into the environment is less understood and difficult to quantify. The impact on sea mammals of the radiation of noise into the environment has been identified as one of the main causes of their decline. The noise being radiated by the vessel in to the environment can be divided into three categories. The first is the interior machinery noise which is transmitted through the hull into the water. The second is the hull flow noise over the hull including the bow wave and stern wave. The third component is the noise associated with the propeller(s). The focus of this paper is to develop a method using computational fluid dynamics to predict the hull flow noise and propeller noise. The analysis can be done relatively early on in the design stage to compare hull forms, appendage arrangements and propeller designs.

2. ANALYSIS METHOD & APPROACH

The initial assumption is the noise can be broken into components to be added together at the end. The hydrodynamic noise is broken into three components. The components are based on their frequency ranges. The first is the flow noise over the hull where the frequency range is up to 165 hertz. The second component is the steady state propeller loading with a frequency range up to 1000 hertz. The final component is the propeller nonlinear component related to cavitation. The frequency range associated with cavitation in our example is up to 3000 hertz. The range of 3000 hertz is limited by the time step in the CFD analysis. In the case where the cavitation levels are higher the spectrum would extent to 10 kHz.

2.1 APPROACH

In order to predict the three noise components, three CFD models were developed at full scale. The model to be analysed is an existing double ended ferry, Figure (1). The principal dimensions of the ferry are outlined below.

LOA	140 m
LWL	134.4 m
BOA	27.4 m
BWL	23.24 m
Δ	5335 mt
C_b	0.346
T	5.18 m
Power	(4) 3000 kW Generators

Figure (1): 140m Ferry

The first CFD model incorporated the hull to the main deck with all appendages to predict the resistance, bow wave and stern wave components of the flow noise. In each model a series of fifteen probes were in the model. The probes were in three groups of five over the length of the vessel, a set at midship and at the bow and stern propellers. The probes were located at a radius of 20m perpendicular to centreline at the waterline. The radius was clocked from the horizontal at 15, 45 and 90 degrees on each side of the model for a total of five probes at each of the three sections. The layout of the probes is shown in Figure (2).

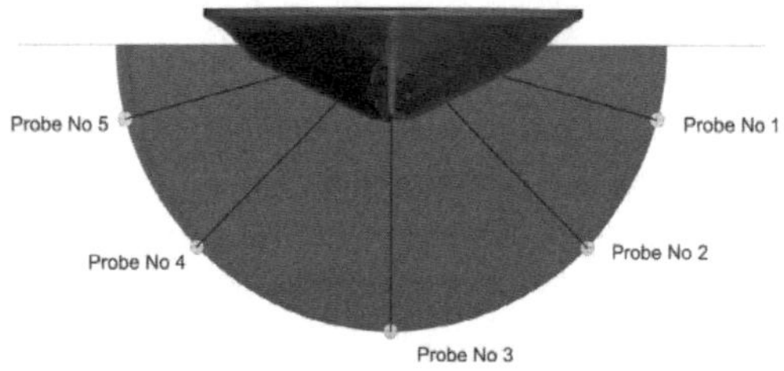

Figure (2): Probe Locations at each section

The position of the probes was the same for each model. The overall domain of each of the CFD models remained the same for each model. Each of the models has a cell count of slightly over 60 million cells. The hull model was run on a cluster of 336 cores with a nominal time step of 0.003 seconds. At this time step the frequency range is limited to 165 hertz.

The steady state propeller model incorporates the two propellers, one at each end of the vessel. Figure (3) shows the propeller and rudder arrangement. Under normal operation of the vessel, the stern propeller provides 90 percent of the thrust while the bow develops up to ten percent. The propeller models include two independent meshes in addition to the overall mesh. The two rotating domains are at each end of the vessel. The aft propeller domain rotates at 19.9 radians per second while the bow propeller rotates in the opposite direction at 15.9 radians per second. The propeller shaft has a rake of 2.8 degrees which requires the rotating domain set at the same rake angle.

The steady state propeller model runs on 168 cores with a nominal time step of 0.0005 seconds. The sample rate limits the frequency range to 1000 hertz.

Figure (3): Propeller Arrangement

The third model uses the propeller model and adds a second multiphase component with the addition of water vapour. The cavitation model has both a free surface with air as well as the addition of water vapour. The model was run on a cluster of 224 cores at a nominal time step of 0.00016 seconds. The sample rate limits the frequency range to 3125 hertz.

2.2 ANALYSIS METHOD

The model provides a range of pressure data at each probe for each time step. The pressure data is recorded at each time step. Each of the three models was run for two stages. The first stage allowed the model to come up to a relatively steady state. The second stage provided a range of data that provided a steady mean loading for analysis of the unsteady loading generating pressure pulses. The entire second stage was used in the analysis of the noise level. The length of time for the two stages was the same.

The data from each pressure probe was run through a FFT program to determine the excitation in the frequency domain. In order to focus in on the dominant frequencies, the Power Spectral Density (PSD) was used to estimate the noise level at each frequency. The average of the excitation was done for each probe and the noise was estimated based on the following formula.

$$SPL = 10Log_{10}\left(\frac{PSD}{p_0^2}\right)$$

Where:

SPL Sound Pressure Level in Db

PSD Power Spectral Density at Frequency f

p_0: reference sound pressure = 1μPa

Each model provides a range of frequency data for review. The hull was run at 19 knots resulting in the cavitation model dominating the results. The results were compared to noise data models for similar ships and good correlation was found except at low frequency for the propeller cavitation component. The cavitation seems to increase the lower frequency excitation below 100 hertz as well as add the higher frequency components.

3. ANALYSIS RESULTS

The analysis results are provided for each model to define the level and range of noise associated with each component. Each model was run through the same analysis procedure to estimate the narrow band noise levels.

3.1 HULL FLOW NOISE

The hull flow noise includes the pressure fluctuations generated by flow over the hull, the free surface and appendages. The hull model does not include the normal discontinuity of the hull surface over frames, hull anodes, keel coolers, sea chests, etc. The resulting peak pressure is at frequencies below 10 hertz. These high energy excitations may generate strike tones on the hull structure that will generate noise at higher frequencies like a hammer ringing a bell.

The mean pressure between the probes at the three positions along the length of the hull is consistent with what is expected. The hull pressure over the bow probes is positive with a mean pressure of approximately 3500 Pa. The pressure over the probes at midship are negative with a mean pressure of -3150 Pa and the pressure distribution of the probes at the stern have a mean pressure of 600 Pa. The plot of the pressure for each probe is provided in Figures (4) through (6).

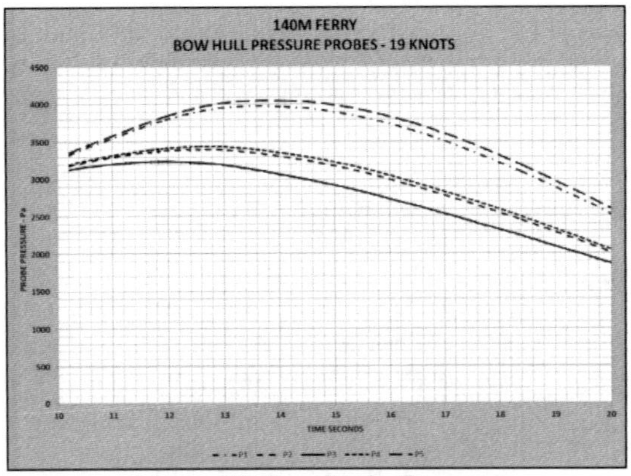

Figure (4): Bow Probe Pressures

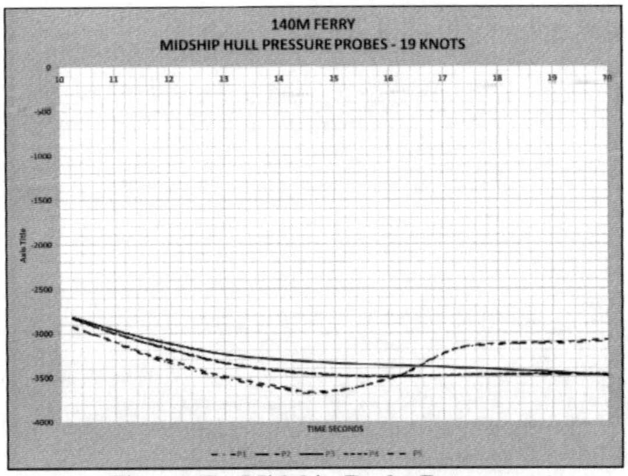

Figure (5): Midship Probe Pressures

The stern pressure probes exhibit a positive pressure in the hull model in the absence of the propeller thrust, Figure (6). The stern probes in the propeller models show a negative pressure relative to the ambient pressure providing insight into the thrust deduction and impact of the propeller wash in the hull resistance.

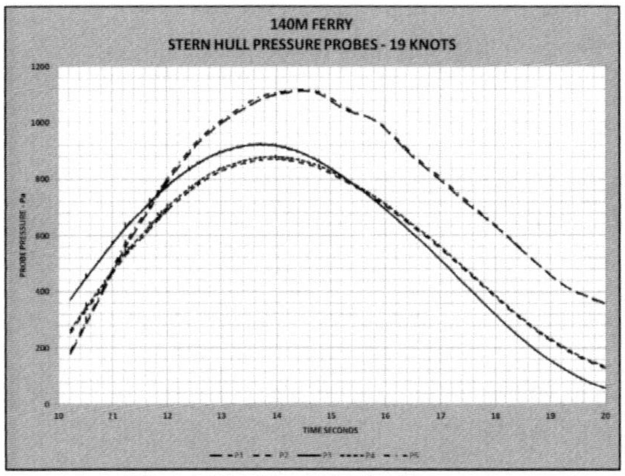

Figure (6): Stern Probe Pressures

The pressure data for each probe was evaluated through a Fast Fourier Transform (FFT) program to determine the frequency domain of the data. The peak response at each of the three positions along the hull was analysed at probes (1) and (5). For clarity, this point was used to present the frequency domain of the data and was used to determine the noise level at the position along the hull.

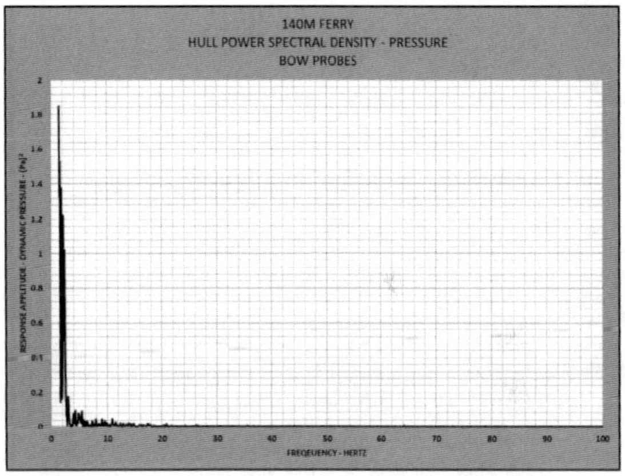

Figure (7): Bow Probe Pressure Amplitude vs Frequency

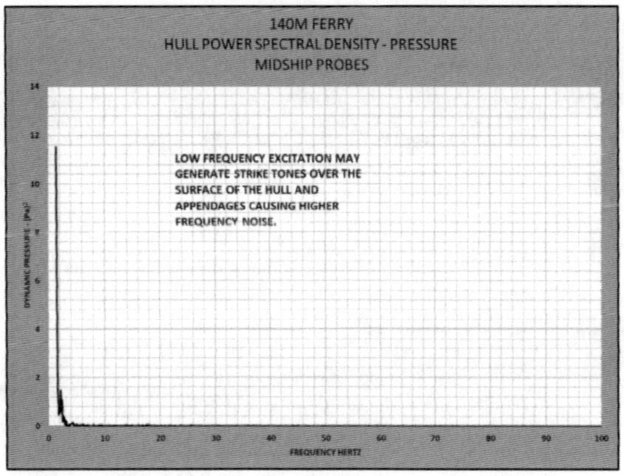

Figure (8): Midship Probe Pressure Amplitude vs Frequency

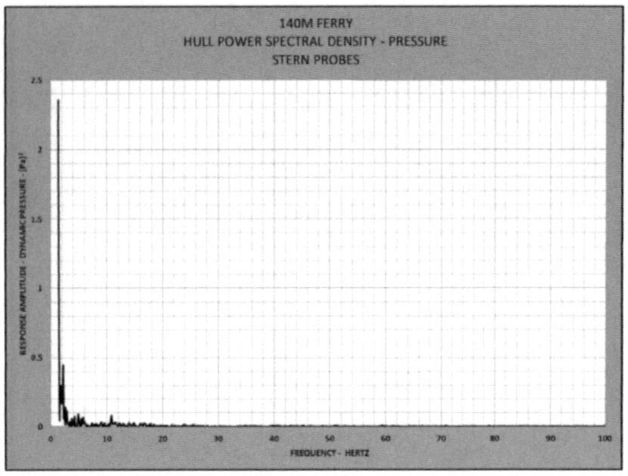

Figure (9): Stern Probe Pressure Amplitude vs Frequency

The frequency plots for each position along the hull are shown in Figures (7) through (9). Figures (10) through (12) show the noise level in dB (1μPa) at 1m at the bow, midship and stern respectively.

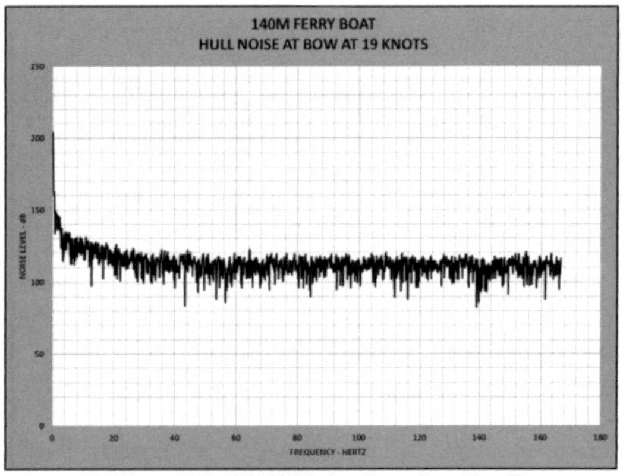

Figure (10): Hull Noise Level at the Bow

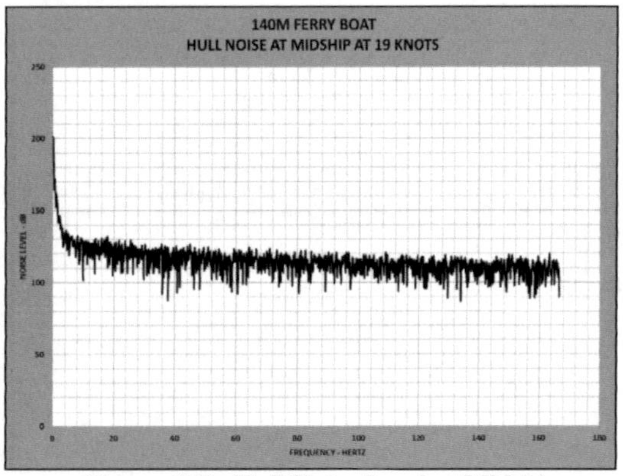

Figure (11): Hull Noise Level at Midship

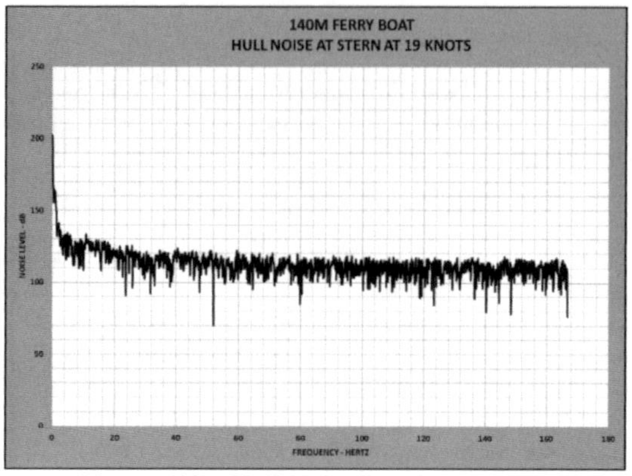

Figure (12): Hull Noise Level at the Stern

In addition to the noise level the CFD model results also provided resistance data, trim and sinkage of the hull at the design displacement condition. The data results are shown in Figure (13) versus time over the second stage run.

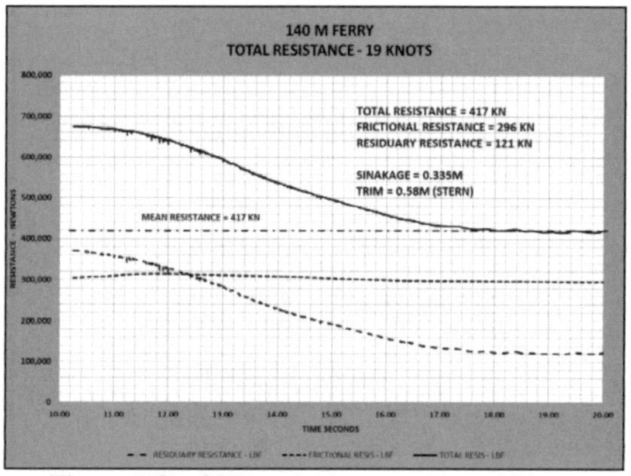

Figure (13): CFD resistance results at 19 knots

Figure (14) shows the CFD model running on the surface provides qualitative information on the bow wave and quarter wave.

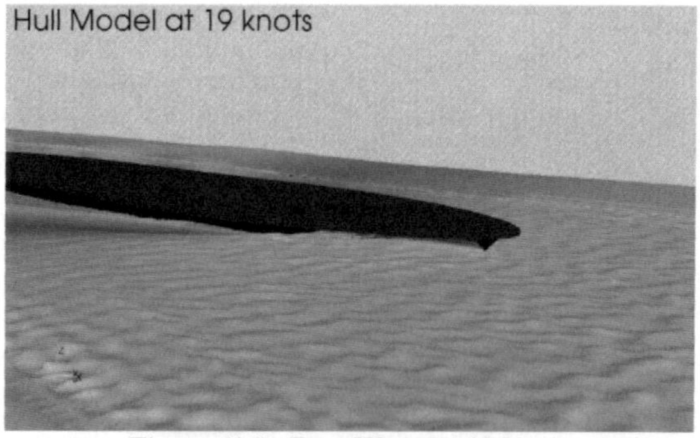

Figure (14): Bow Wave at 19 knots

3.2 PROPELLER NOISE NON-CAVITATING

The bow and stern propellers were added to the hull model along with the rotating domains. The model was run in two stages to allow the model to reach a steady mathematical equilibrium in an initial run. The second run was a restart of the first run allowing a steadier load on the hull and propellers.

The probes are in the same location and the data is used in the same manner although the time step is shorter providing a wider range of frequency data. Figure (15) shows the detail of the propeller in the CFD model.

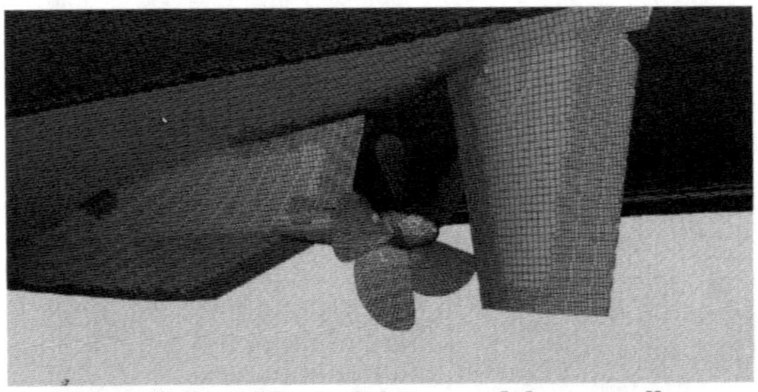

Figure (15): CFD mesh in way of the propellers.

The pressure data at each propeller probe is plotted in Figures (16) and (17). The data shows a clear alternating component.

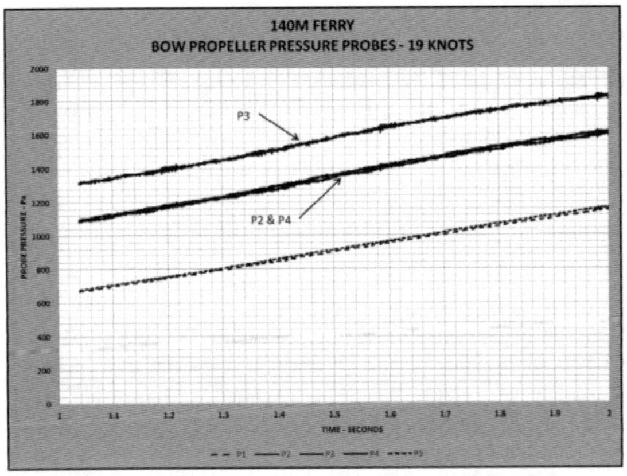

Figure (16): Bow Propeller Pressure Probes

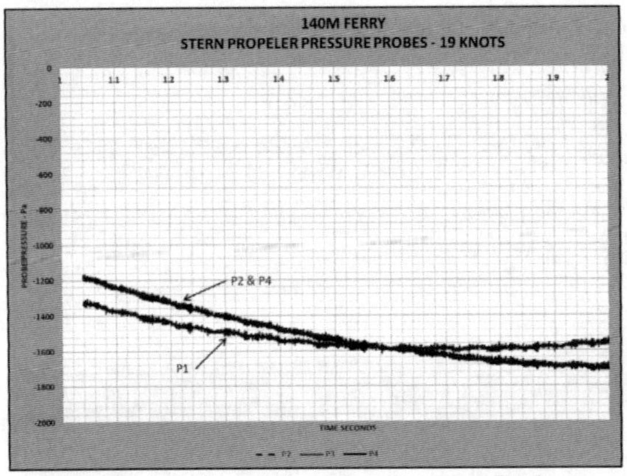

Figure (17): Stern Propeller Pressure Probes

The peak excitation appears in probe number 3 in way of each propeller. Probe number 3 is at bottom centreline approximately 17 meters below the centreline of the propeller hub.

The contribution of the propellers to the midship probes was shown to be two orders of magnitude below the amplitudes found at the propellers. The result was to be expected, but it helps validate the methodology.

The FFT of the stern and bow propeller results are shown in Figures (18) and (19). The plots show a wider range of excitation frequencies as high at 650 hertz. The highest response on the stern propeller over the whole range was from the No 3 probe located below the propeller. The highest response on the bow propeller was split between the No 2 probe for the low frequencies below 50 hertz and the No. 3 probe for the higher frequencies. Both plots clearly show the blade rates at 12.7 hertz for the bow propeller and 15.8 hertz for the stern propeller.

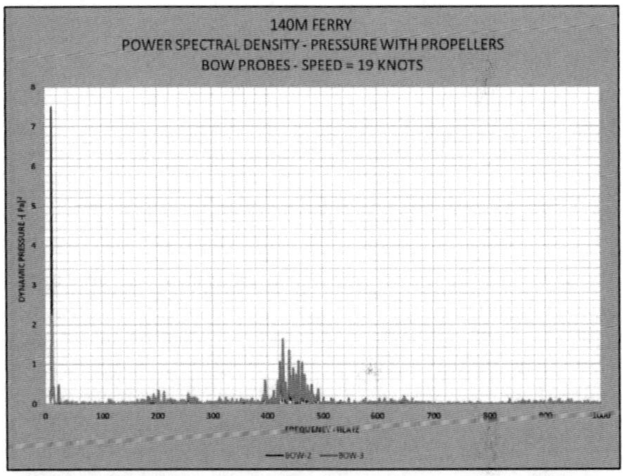

Figure (18): Bow Propeller Power Spectral Density

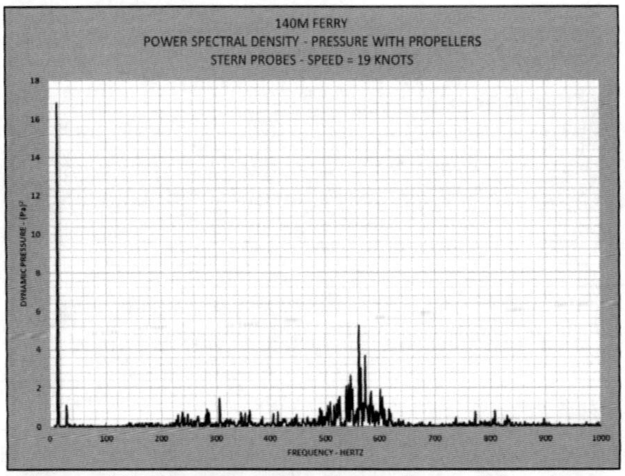

Figure (19): Stern Propeller Power Spectral Density

The results in decibels are presented in Figure (20) showing the propeller non-cavitating impact up to 1000 hertz.

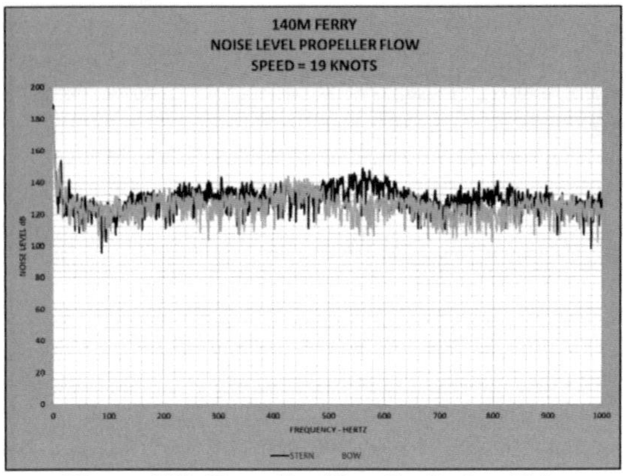

Figure (20): Propeller Non-Cavitating Noise Spectrum

The propeller loading for the bow and stern propeller show good agreement with the propeller curve and design data. The efficiency of the propeller shows excellent agreement with the open water value. The full data is provided below.

Stern Propeller

$K_t = 0.1986$
$K_q = 0.0328$
$J = 0.686$
$\eta = 0.661 \quad (1\text{-}\omega) = 0.88$

The bow propeller is operating in the third propeller quadrant under a substantial wake caused by the forward rudder. Figure (21) shows the large alternating thrust and torque on the propeller.

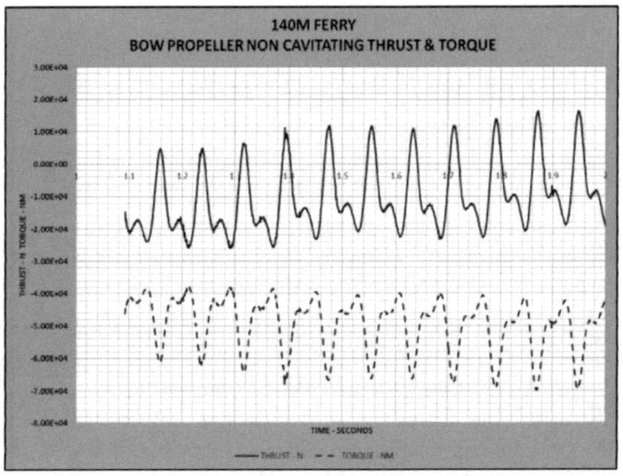

Figure (21): Bow Propeller Thrust & Torque

The hull model provided the propeller wakes for both the bow and stern propellers. The stern propeller wake is typical for the hull form and principal dimensions. The bow wake is more dramatic behind a large rudder and bow wave. The wake has a hole at the upper 0.70 propeller radius resulting in a large gust load on the propeller blades as they pass through it. Figure (22) shows the wake of the bow propeller at 19 knots.

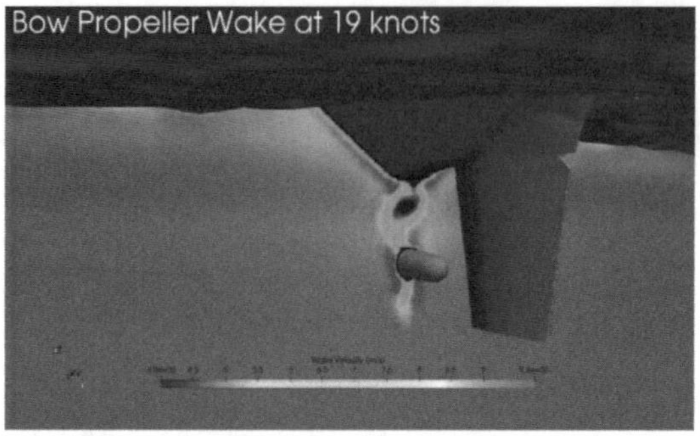

Figure (22): Bow Propeller Wake at 19 knots

Figure (23) shows the stern propeller wake at 19 knots. The average wake is estimated at 0.880. In addition the propeller shaft has a rake of 2.8 degrees.

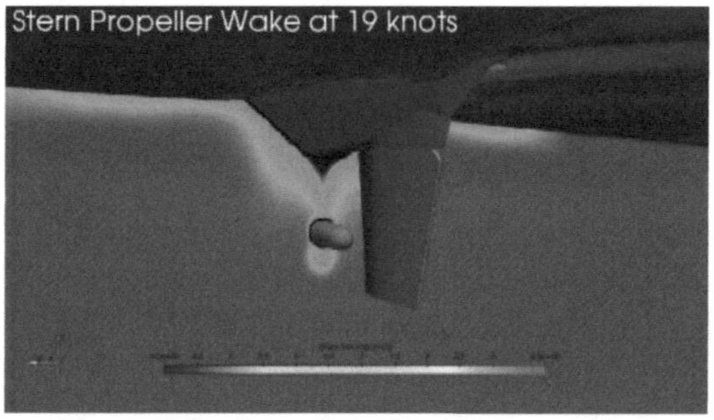

Figure (23): Stern Propeller Wake at 19 knots

3.3 PROPELLER NOISE CAVITATING

The vessel has a design top speed of 21 knots with (4) generators running. The speed of 19 knots is the maximum speed with (3) generators operating. The speed of 19 knots is the top speed of the normal operating range of the boat. The level of cavitation is less than five percent on the stern propeller based on the Burrill Diagram.

The propeller responses were consistent with the non-cavitating model analysis. The cavitating model propeller produced more thrust and more torque due to an increase in the wake at the stern propeller. Figures (24) and (25) show the cavitation cavity on the back and face of the stern propeller.

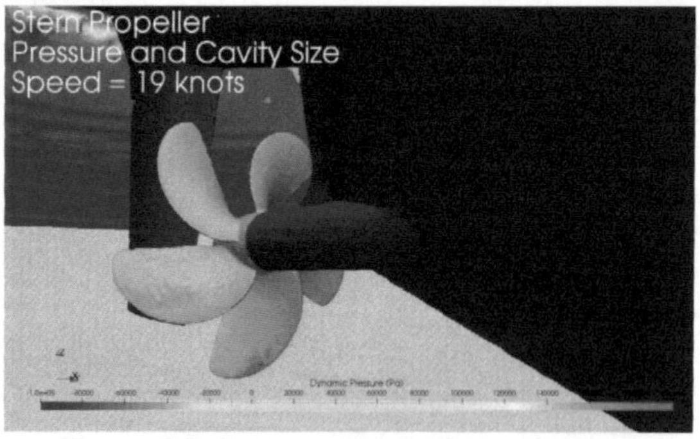

Figure (24): Pressure and Cavity on the Stern Propeller

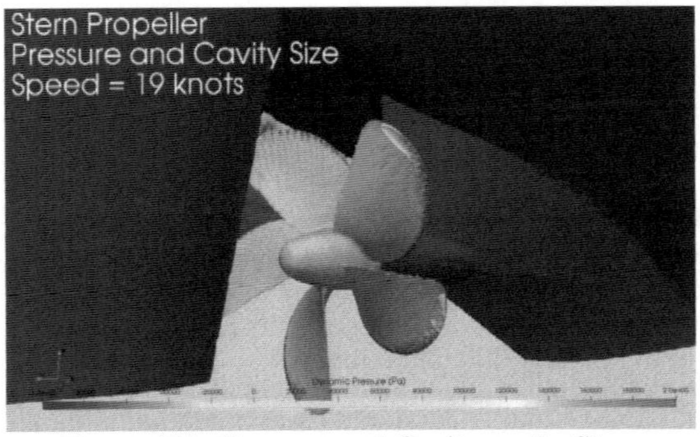

Figure (25): Pressure and Cavity on the Stern Propeller

The thrust and torque response are provided in Figure (26). The excitation is an order of magnitude higher than the non-cavitating model. Considering the low level of cavitation, the model may be overstating the noise levels.

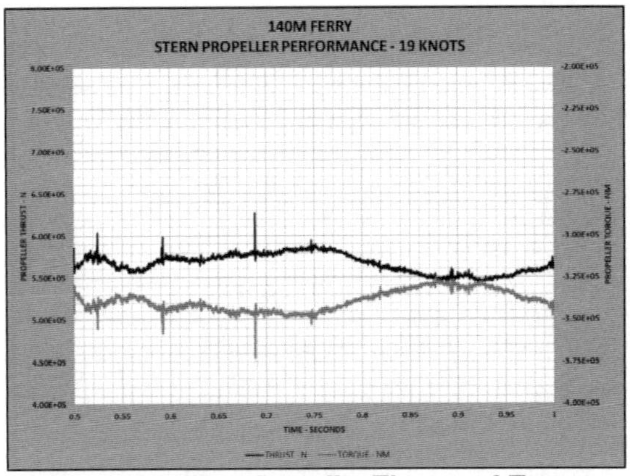

Figure (26): Stern Propeller Thrust and Torque results

Stern Propeller with cavitation:

$K_t = 0.2230$
$K_q = 0.0339$
$J = 0.686$
$\eta = 0.718 \quad (1\text{-}\omega) = 0.88$

The bow propeller is lightly loaded and is generating some power back into the system. Figures (27) and (28) illustrate the vapour cavity on the face and back of the bow propeller. The figures show a vertical cavity floating between the rudder and the propeller. The cavity is assumed to be the result of the wake at the 0.70 radius caused by the rudder and bow wave.

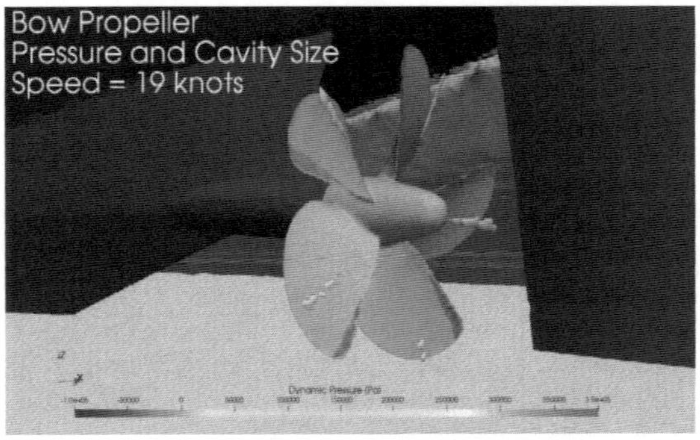

Figure (27): Bow Propeller Face Cavity

Figure (27) shows the floating cavities as a result of the flow off the leading edge of the rudder while running as the bow. The wake impact generates a large excitation from 100 to 1000 hertz.

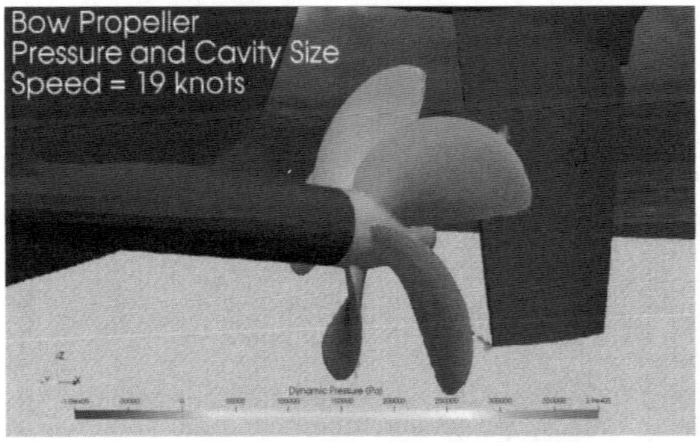

Figure (28): Bow Propeller Back Cavity

The probe pressure plots are provided in Figures (29) and (30). The bow pressure probes show the blade pass at 0.079 seconds as the inverse to the blade rate of 12.7 hertz.

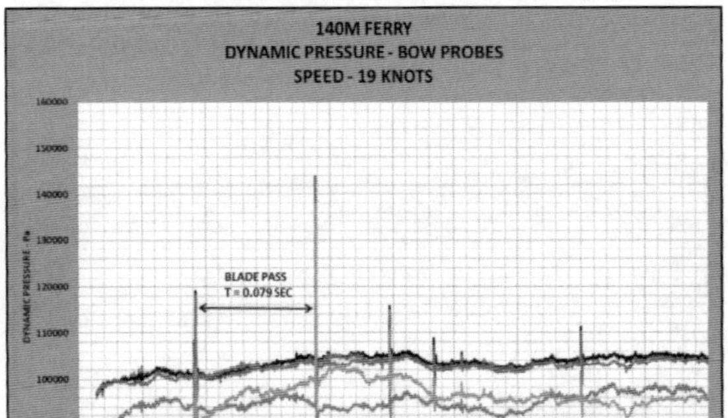

Figure (29): Bow propeller pressure probes

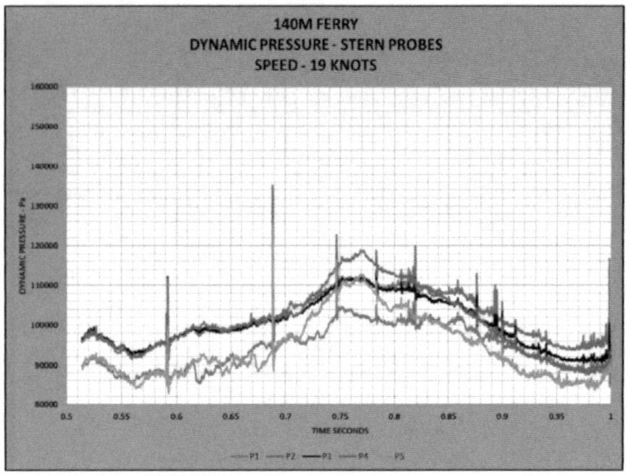

Figure (30): Stern propeller pressure probes

The frequency plots for the bow and stern propeller are shown in Figures (31) and (32). The amplitudes shown are the peak amplitudes determined in the FFT analysis. The transformation of the data to decibels uses the PSD value at each frequency averaged over the five probes.

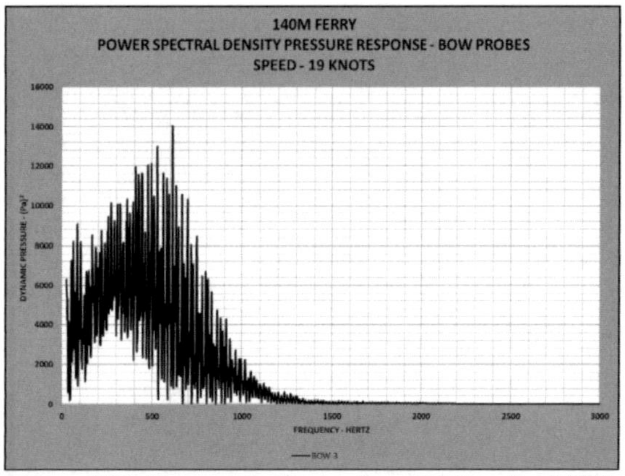

Figure (31): Bow probe Power Spectral Density (PSD)

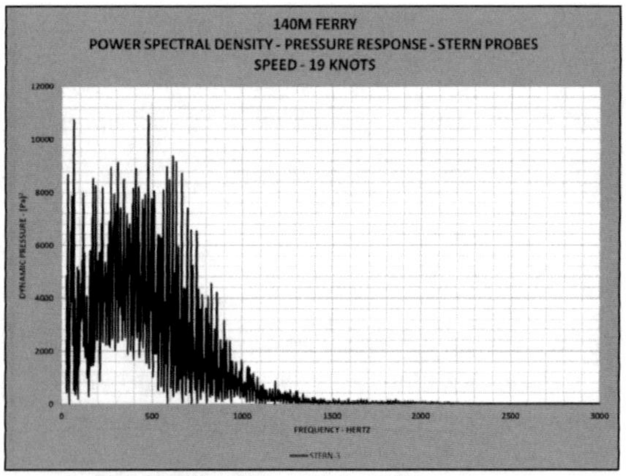

Figure (32): Stern probe PSD response

The noise prediction for the cavitating propellers is provided in Figure (33). The plot is a narrow band plot of noise dB (1μPa) at 1 meter radius.

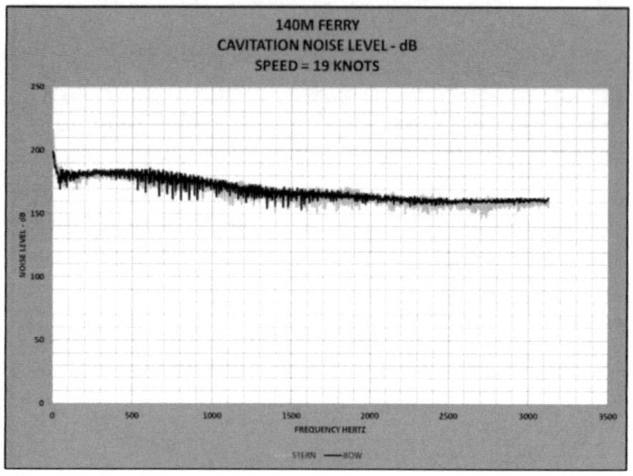

Figure (33): Noise Spectrum for Cavitating Propellers

4. CONCLUSIONS AND FUTURE WORK

The analysis process included three separate CFD models. The models differentiated the three primary sources for hydrodynamic noise over the hull of a large vessel. The three components are hydrodynamic flow noise, non-cavitating propeller noise and cavitating propeller noise. The cavitating propeller noise dominates the noise level in the example calculation shown above even at a relatively low cavitation level. Each analysis utilized a large CFD model. The refined models were a good representation of the ship providing detail data for the hydrodynamic performance of the vessel regarding propeller performance, vibration and hull resistance. The noise levels are valuable tools for comparison and as part of the development of a noise and vibration inventory. The models can be used effectively for wake studies and propeller design validation. It is clear in the application that the bow propeller wake is a design problem where some alternatives could be investigated.

The noise signature of a ship includes other noise components which need to be included into the analysis. The machinery noise operating in the vessel radiates noise over a broad frequency range. The main propulsion system dominates the radiated noise spectrum with a bandwidth extending to 10k hertz at lower speeds where cavitation is eliminated.

Another feature of the hydrodynamic noise estimate is that calculated excitation is only an estimate of the pressure pulse or strike tone on the hull surface. The steel hull will then generate a strike tone depending on the ring frequency of the particular area of the hull. The strike tone is assumed to be the tertiary panels of flat plate on the hull. These panels are exposed to the water flow and would be the initial strike surface of any pressure pulse. Figure (34) shows the fundamental frequency of a range of steel plates based on their panel width and thickness.

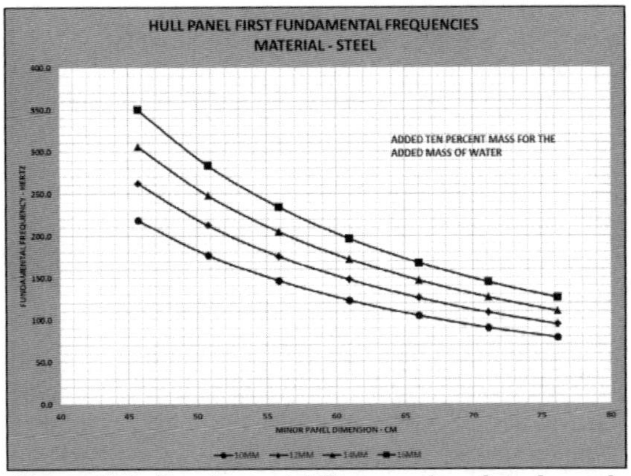

Figure (34): Fundamental Frequencies of steel panels

The hull structure will respond to the pressure pulse or strike and radiate noise at the fundamental frequency and at all the higher modes as well. The higher frequencies will be damped, but since the excitation is on the water side of plate the transmission loss will be minimized depending on the geometry of the plate. The geometry considerations include aspect ratio and curvature of the individual panels.

The internal noise in the machinery spaces on board the vessel radiate noise through the steel hull into the water. The transmission efficiency through steel to water is approximately 0.141 compared to 0.00004 through steel to air [5]. Most machinery spaces are large with mostly hard surfaces. Noise attenuation has been installed between the manned spaces and the machinery space, but very little has been done over the portion of the hull that is in contact the water.

Several methods have been proposed to estimate the expected noise level of the vessel. The models are generally broken into components.

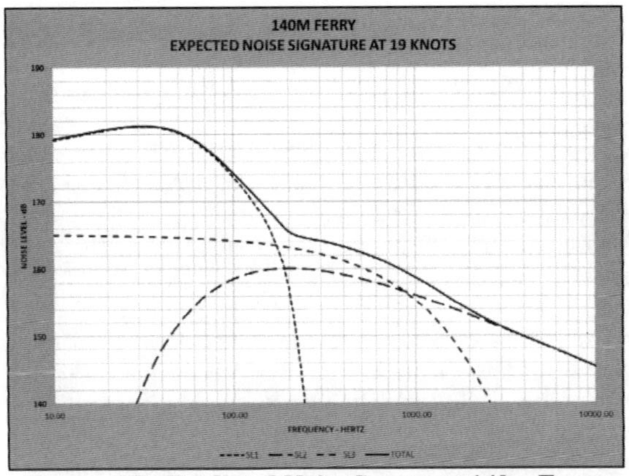

Figure (35): Predicted Noise Spectrum 140m Ferry

Reference [4] proposed a method based on a large sample of ship noise data. Figure (35) provides a plot of the expected noise level of the 140m ferry modelled above at 19 knots.

The spectrum is broken into three components. The first is the low frequency propeller noise. The second is the high frequency propeller noise and the third component is the machinery noise. The machinery noise was deducted from the predicted noise level and provided as Figure (36). Figure (37) shows the results of the analysis with the non-cavitating and cavitating components of the propeller spectrum.

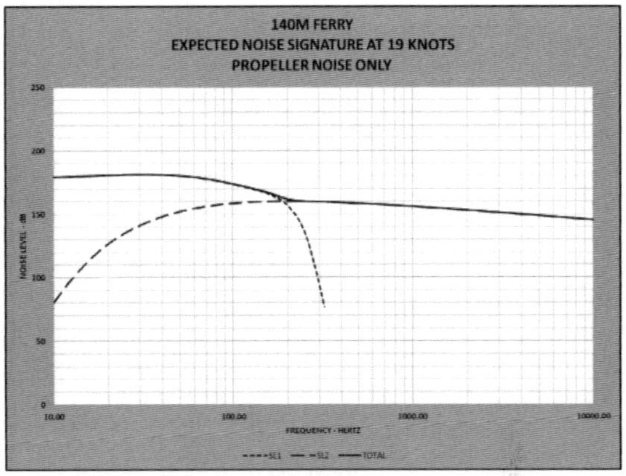

Figure (36): Predicted Noise Spectrum w/o Machinery

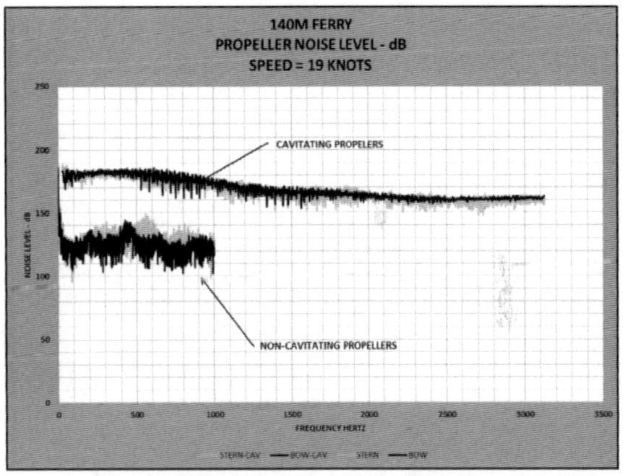

Figure (37): Propeller Noise Spectrums

The propeller related loading is in good agreement with the baseline data provided in reference [4].

The predicted value based on the range of vessel data may be lower based on ocean going vessels with a single propeller. In addition, the ferry is running at a relatively high speed for the waterline length compared to ocean going vessels.

The rake of the propeller also has an impact. The rake causes a wake on each side of the propeller where the rake increases the angle of attack on the starboard side and decreases it on the port side for a right handed propeller. The cavitation cavities are all on the starboard side with the largest cavity at the 3:00 o'clock position. Figure (38) shows the dramatic difference in the cavitation cavities on each of the propeller blades.

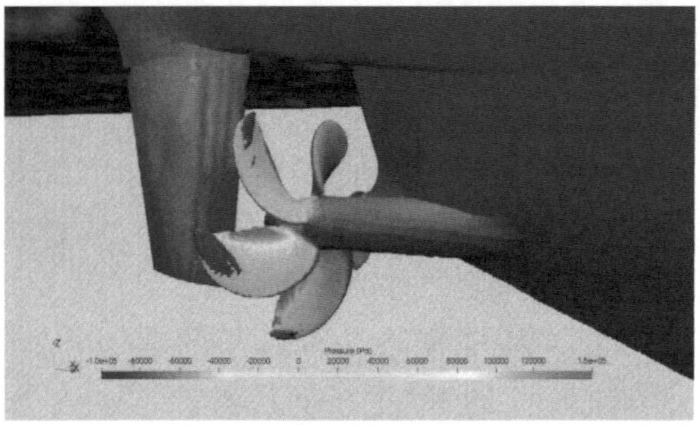

Figure (38): Stern Propeller location of Cavitation Cavities

Actual vessel data is shown in Figure (39), reference [7]. The vessel data is provided at relatively low speed up to 12.6 knots. The data was taken close to Coleman Dock in Seattle where the vessel lands. The data speed range was from 8.5 to 12.6 knots. The noise rate of change versus speed showed a reverse slope in the narrow speed range. The assumption is the vessel noise levels are dominated by internal machinery noise at these speeds especially while slowing down on the approach to the terminal. The generators are synchronous and running at a constant 900 RPM.

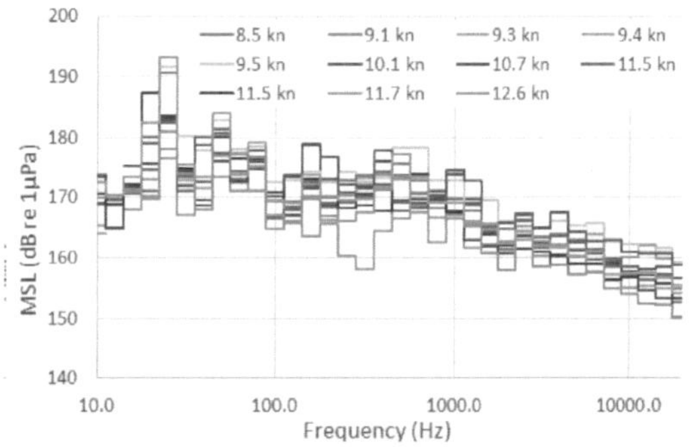

Figure (39): Vessel Data at Speeds up to 12.6k

There is a plan in place to collect more noise data at higher speeds that would include different routes and vessels in the fleet. The normal operating speed for the 140m ferry is 18 knots with a schedule makeup speed of 19+ knots.

When reviewing the final data, the noise levels need to be adjusted to the auditory frequency ranges of different marine mammals. Reference [8] presents a scale to adjust the noise levels to better estimate how each specie will respond. Figure (40) provides a plot of the Low-Frequency Cetaceans M-weights. There are (3) curves provided in reference [9] covering low, medium and high frequency hearing cetaceans.

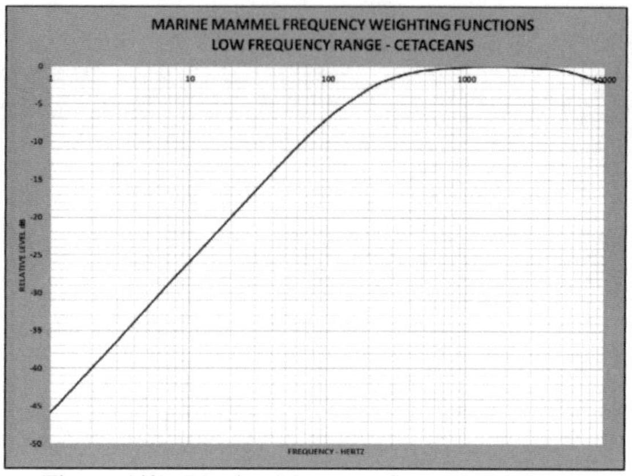

Figure (40): Marine Mammal Sound Weighting Function

The noise spectrum analysed for the hull and appendages is relevant for military and research vessels. The propeller noise for any vessel in transit would dominate the noise spectrum. The hull noise spectrum is dominant at low speed and at low frequency where self-noise and detection would be important. Figure (41) shows the hull flow noise spectrum.

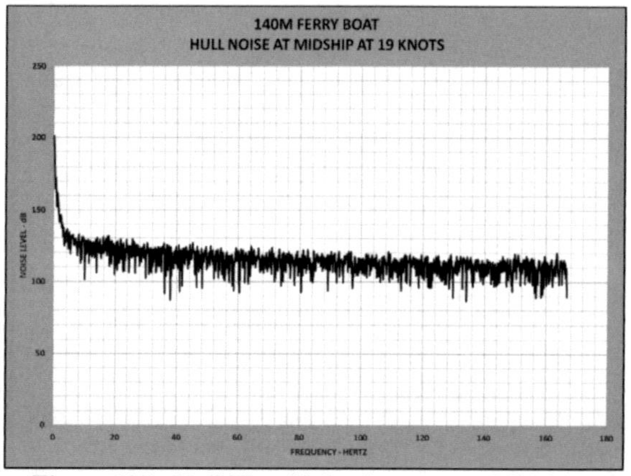

Figure (41): Hull Flow Noise Spectrum - Midship

The hull skin has a roughness specification in OpenFOAM to provide an estimate of a painted surface. The model surfaces are the faired lines of the hull without any discontinuities such as weld distortion, sea chests, keel coolers or anodes. The hull panels will all have strike tones that will add three or four modes based on the modal density of the smaller panels. The modal density of larger sections is lower with separation between ten and twenty hertz. The small panel will have modes separated by 200 to 1000 hertz.

The analysis presented provides a reliable estimate of the hull and propeller components of the noise spectrum for the vessel. Figure (33) shows the total noise level including all the hydrodynamic noise components. The hull model and non-cavitating models provide insight into the noise components and associated wake of the propellers. The data is available at a wide range of vessel conditions and speeds for comparison. In addition, the analysis provides a great deal of data on the hull and propeller performance to the naval architect and marine engineer.

As the methodology and software continue to improve, a reliable and robust procedure will become available to optimize the speed of the vessel while mitigating the impact to the environment. The sound sources discussed above were all considered to be monopoles due to the size of the model and relative scale of the propellers to the length of the hull. In the case of the cavitating propeller model, future work would look at different models and probe positions to possibly look at the cavitating propeller blades as a group of independent monopoles or as a more complex multipole object. The model would relate the interference of each blade in the radiation of pressure pulses. The propeller modelled with the hull and appendages is critical to the accuracy of the performance and noise level. The Mach number in the cavitating model is high enough so the noise level is not overly affected by the analysis model. The propeller in the present analysis is assumed to act as one complex source. The analysis of full scale data may shed some light on the cavitation model and its impact on low frequency noise. The higher noise levels at low frequencies may be an accurate model of the propeller loading due to the change in wake caused by the cavitation. The propeller torque and thrust are higher in the cavitation model compared to the non-cavitating model. The cavitation seems to increase the wake slightly, while increasing the average thrust and holding torque relatively steady.

While research is ongoing to improve the vessel's radiated noise levels, the impact to marine mammals is not well understood and regulatory bodies are beginning to suggest additional requirements to mitigate the radiated noise levels of sea going vessels. The lower frequency noise levels are a major unknown due to the distance they can travel and the addition of those frequencies on the general background noise along coasts in relatively deep and exposed waters.

5. REFERENCES

1. OpenFOAM Users Guide, Version 5.0, OpenFOAM Foundation 24 July 2017,

2. Wilcox, David, Turbulence Modeling for CFD, 3rd Edition, DCW Industries, 2010.

3. Lidtke, Arthur K., Predicting Radiated Noise of Marine Propellers using Acoustic Analogies and Hybrid Eulerian-Langrangian Cavitation Models, PhD Thesis, University of Southampton, June 2017.

4. Wittekind, Dietrich, A Simple Model for the Underwater Noise Source Level of Ships, Journal of Ship Production and Design, Volume 30, No. 1, February 2014.

5. Beranek, Leo, Ver, Istvan, Noise and Vibration Control Engineering, John Wiley & Sons, 1992.

6. MEPC.1/Circ.833, Guidelines for the Reduction of Underwater Noise from Commercial Shipping to Address Adverse Impacts on Marine Life, International Maritime Organization, 7 April 2014.

7. Liefvendahl, Mattias, Feymark, Andreas, Bensow, Rickard, Methodology for Noise Source Modeling and its Application to Baltic Sea Shipping., Deliverable D4.1 of the BONUS SHEBA Project, Chalmers University of Technology, Technical Report 2015:161.

8. Wladichuk, J., Hannay, D., Zizheng, L. WSF Vessel Source Level Analysis, Seattle Terminal, Jasco Applied Sciences, Document 01651, 9 October 2018.

9. Finneran, J.J., Auditory Weighting Functions and TTS/PTS Exposure functions for Marine Mammals Exposed to Underwater Noise. Technical Report for Space and Naval Warfare Systems Centre Pacific, San Diego, 2016

6. AUTHORS BIOGRAPHY

Endicott M. Fay holds the current position of Chief Naval Architect and Manager of Vessel Design at Washington State Ferries. He is a Chartered Naval Architect and Professional Engineer, registered in Washington State. He is responsible for the long-term maintenance, upgrade and preservation of a fleet of 23 car ferries. In addition, Mr. Fay provides consulting in computational fluid dynamics to a wide range of clients in the marine and process industries. Mr. Fay has performed numerous computational projects using the Super Computer Centre at Ohio State University.

NUMERICAL STUDY OF ACOUSTIC CHARACTERISTICS OF A DTMB 4119 PROPELLER DUE TO TIP RAKE

Danio Joe, **Vijit Misra** and **Dr R Vijayakumar**, Indian Institute of Technology Madras, India

SUMMARY

The impact of increased Underwater Radiated Noise (URN) over the past two decades on marine mammals has resulted in the pressing requirement to reduce it. Shipping contributes immensely to the URN. Propeller noise is a major source of URN. The reduction in Propeller noise can hence significantly help in the reduction of URN. With the sole objective of improving the hydrodynamic performance of propellers ways to prevent cavitation are being developed. However, the reduction of non cavitating noise produced by the propeller would still remain a challenge. The change in the propeller geometry can modify the acoustic characteristics. In this present study, effect of modifying the tip of DTMB4119 propeller on the acoustic and hydrodynamic characteristics is presented. The change in the flow pattern at the tip due to introduction of tip rake is also discussed. The SPL has been calculated by using the two-step Ffowcs William and Hawkings (FW-H) equations from the pressure distribution at various points around the propeller. SPL at various points in the downstream and propeller disk plane are numerically predicted and discussed.

NOMENCLATURE

A_C	mean cavitation area on the blades (m^2)
A_D	propeller disk area (m^2)
V_{Tip}	blade tip velocity ($m\ s^{-1}$)
D	Propeller diameter (m)
F	frequency (Hz)
P	Pressure ($N\ m^{-2}$)
c	Speed of sound in water ($m\ s^{-1}$)
L_S	Sound Pressure Level (dB ref 1μPa))
n	Propeller rpm (rpm)
Z	Number of propeller blades
FWH	Ffowcs-Williams Hawkings
V	Ship speed ($m\ s^{-1}$)
V_A	Speed of advance ($m\ s^{-1}$)
URN	Underwater Radiated Noise
υ	Kinematic viscosity ($N\ s\ m^{-2}$)
ρ	Density of water ($kg\ m^{-3}$)
SPL	Sound Pressure Level (dB)

1. INTRODUCTION

Propeller noise is a major contributor to URN. Increase of shipping noise has resulted in an adverse impact on marine life. Reduction of propeller noise also finds its application in producing silent warships with increased acoustic stealth. Hence, efforts to reduce the propeller noise should be made at design stage itself.

The drag is overcome by the thrust produced by the propeller. The thrust producing mechanism of the propeller inherently produces noise. The Propeller faces unsteady forces during its rotation. Pressure waves are created in water during its rotation through four mechanisms. These are (Carlton, 2007):

(a) Thickness noise - The displacement of the water by the propeller blade profile displaces the water in its path of rotation.
(b) Loading noise – During the rotation there is a pressure difference between suction and pressure surfaces of the propeller.
(c) The rotation of blades in variable wake field behind the vessel produces fluctuation of cavity volumes.
(d) The collapse of a cavitation bubble or vortex.

Cavitation is the cause of (c) and (d). (a) and (b) can occur both in cavitating or non-cavitating state. The noise generating mechanisms for a propeller operating in non-cavitation regime consists of broadband noise (due to laminar boundary layer instability noise or vortex shredding, inflow turbulence and surface turbulence) (Richard and Munjal, 1980)(Parchen, 2000) and tonals (due to steady and fluctuating component of thrust and torque and thickness effects). The modification of the vortex shedding can therefore affect the propeller noise.

Tip vortices are an unavoidable side effect of producing lift with an aerofoil of finite length. An aerofoil is designed first to have relatively high pressure at the face and lower pressure at the back. Second, fluid passing over the top of the aerofoil is directed downwards as it leaves the trailing edge (thus "pulling" the wing upwards by Newton's Third Law). These two characteristics of an aerofoil combine to make fluid around the bladetip to "spin". High-pressure fluid below the blade "leak" upwards over the bladetip, then be directed back downwards along with the rest of the fluid, creating the tip vortex.

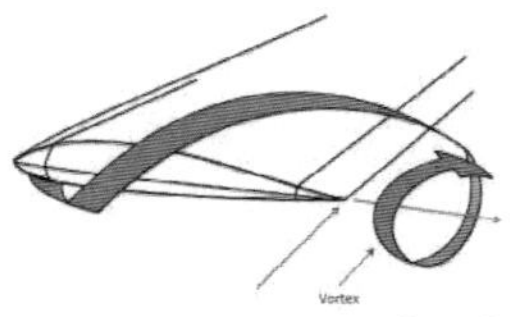

Figure1:Tip vortice on a conventional aircraft wing

Reducing the pressure differential between fluid above and below the blade will by definition reduce lift, because it is this pressure differential producing lift. By the same token, reducing the "downwash" off the back of the blade will reduce lift because the force creating the downwash also produces lift on the blade. Therefore, conventional wings will produce wingtip vortices by definition. A way to reduce vortices is therefore to make it harder for the fluid to "leak" upward over the bladetip. This is the principle behind the "winglets" seen on many aircrafts (Figure1). The idea of tip rake was applied on the airplane wing (Cone and Clarence 1962). They prevent air underneath the wing being able to move over the top of it, thus reducing the severity of the rotation of the vortex (which reduces drag). Modified tip propellers, Kappel propeller and the Contracted and Loaded Tip (CLT) propeller (Gennaro and Gonzalez-Adalid2012; Gaggeroet al. 2016) are available for marine applications. However, in all the past research, focus has only been on improving the hydrodynamic efficiency of the propeller.

The experimental measurement of noise of a cavitating propeller were made by (Arakeri et. al. 1990).The first numerical prediction of propeller noise in cavitating regime was made by Seolet. al. in 2003, wherein he used DTMB4119 model propeller and potential based panel method coupled with acoustic analogy. Kheradmand et al. in 2014 carried out the numerical simulation of tonal and broadband hydrodynamic noises in non-cavitating regime. The RANS model was used for hydroacoustic studies on DTMB 4119 (Hu et al. 2009). Numerical prediction of flow around a propeller was investigated (Li and Yang 2009). The numerical study on effects of introducing tip rake on the acoustic properties of DTMB 4382 propeller with an inherent skew has been carried out(Ghassemi et al, 2018).Further, a study on the introduction of tip rake angle of - 6^0 was also attempted(Danio et al.,2018) This study attempts to take the investigation on the effect of tip rake angle on conventional propellers further. Applying the same technique used to reduce vortices in an aircraft wing, a rake angle is introduced in the tip of a propeller blade. In the present study, the effect this modification has on the propeller efficiency and the SPL is attempted.

2. GOVERNING EQUATION AND METHODOLOGY

The methodology adopted to predict noise from a marine propeller has been used over the years (Tewari et al.,2019).The prediction of noise originating from the presence of a turbulent flow is done using FWH acoustic analogy (Williams *et al.*, 1969). The FWH theory in an indirect method for numerical prediction of noise. Surface pressure sources as well as turbulent flow pressure sources form part of the FWH equation. Being an indirect method the pressures on the surfaces needs to be found. The CFD analysis of the flow is carried out using discretization techniques like FEM, FDM or FVM. The PDEs of the governing equations (1)(2) are used to solve the flow field variables for the set of boundary and initial conditions of the computational domain.

Continuity Equation

$$\frac{D\rho}{Dt} + \nabla(\rho u) = \frac{\partial(\rho u_i)}{\partial x_i} + \frac{\partial(\rho u)}{\partial t} \qquad -(1)$$

Navier Stokes Equation

$$\frac{\partial(\rho u_i)}{\partial t} + \frac{\partial(\rho u_i u_j)}{\partial x_j} = -\frac{\partial p}{\partial x_i} + \frac{\partial \zeta}{\partial x_i} + \rho f_x \qquad -(2)$$

The flow was simulated using CFD software STARCCM+. The blade harmonic noise components can be effectively captured using unsteady RANS method (Lidtke, Turnock and Humphrey, 2015).Hence, RANS solver is used in the hydrodynamic analysis.

2.1 PROPELLER GEOMETRY

DTMB 4119 propeller (Table1) (Figure 2(a)) has been used for the study. CAD models with tip rake(+/- $\mathbf{6^0}$ and +/- $\mathbf{10^0}$) (Figure 2(b)) were created. The models of the propeller were created using Rhino™.

Table 1: Principal characteristics of the 3 bladed DTMB4119 propeller

Parameter	Value
Number of blades, Z	3
Skew (degree)	0
Diameter (m),D	0.305
Section thickness form	NACA 66
Boss ratio	0.2
Design advance ratio	0.833

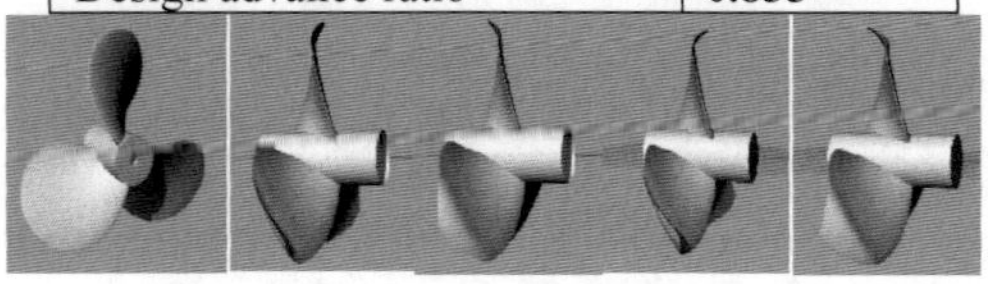

Figure2: (a)DTMB 4119 Propeller (b) Tip rake -/+ 6^0 (c) Tip rake -/+ 10^0

2.2 DOMAIN AND MESHING

The computational domain, Figure 3(a) consists of a rotating part (around the propeller to capture rotation effects) and a stationary part (to capture the flow around rotor). A cylindrical domain has been used domain- (Sezen, Dogrul and Bal, 2016), (Noughabi, Bayati and Tadjfar, 2017). ITTC guidelines are used for domain sizing to nullify effect of the boundaries i.e. no reflection from the boundaries. (ITTC, 2014)Figure 3(a).

A hybrid and unstructured mesh was used for grid generation. Finer meshing has been used at blade tips and edges. A conformal grid has been obtained at the Rotor-Stator interface .The cell size increases progressively from 5% to 500% of the base size. The computational cost is economized without affecting the capture of flow parameters. The base size of the cell is 20mm. About two million cells have been created in the domain, as shown in Figure 3(b)(c). Realizable *k-ε* model for turbulence Modeling are used. The Rotor is given a rotation of n=10 rps. Open water characteristics of DTMB 4119 have been estimated and compared with results from published literature(Bagheri, Seif, *et al.*, 2015), (Sezen, Dogrul and Bal, 2016).

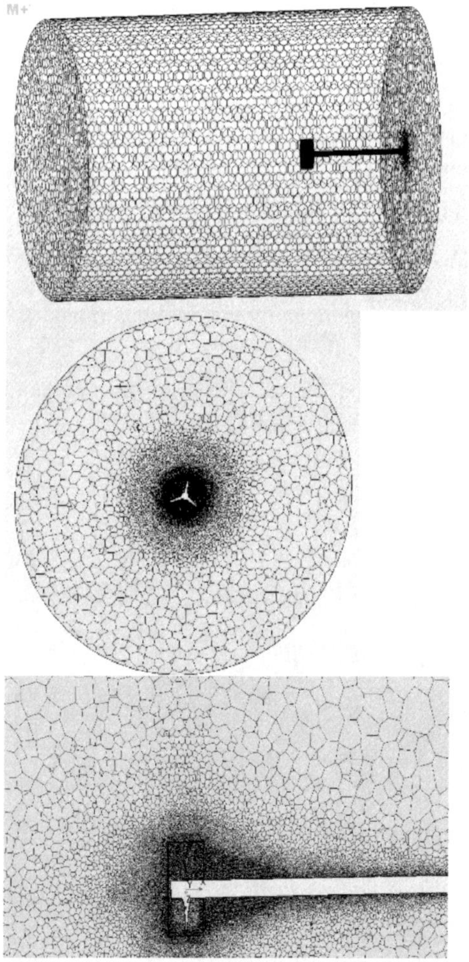

Figure 3: (a) Computational domain (b) Cross section Mesh Diagram (c) Axial mesh diagram

2.3 HYDRO-ACOUSTIC ANALYSIS: FFOWCS-WILLIAMS HAWKINGS EQUATION (FWHE)

FWH acoustic analogy is used to predict noise from a turbulent flow (Williams *et al.*, 1969). The FWH equation includes surface sources (monopole or thickness sources and dipole or loading sources) and quadrupole sources introduced by Caridi (2007). FWH equation is shown below:

$$\frac{\partial^2 p'}{c_0^2 \partial t^2} - \nabla^2 p' = \frac{\partial^2}{\partial x_i x_j}\left[T_{ij} H(f)\right] - \frac{\partial}{\partial x_i}\left(\left[P_{ij} n_j + \rho u_i (u_n - v_n)\right]\delta(f)\right) + \frac{\partial}{\partial t}\left(\left[\rho_0 v_n + \rho(u_n - v_n)\right]\right)\delta(f) \quad \text{-(3)}$$

Where, p' is the source pressure level at the far-field, (p'=p-p$_0$)), c$_0$ is the sound speed at far-field. T$_{ij}$ is the Lighthill stress tensor (Lighthill M., 1952) (Eqn 4).The rest terms at RHS are quadrapole, dipole and monopole sources respectively.

$$T_{ij} = \rho u_i u_j + \delta_{ij}(p - \rho c_0^2) + \tau_{ij} \quad \text{- (4)}$$

The terms are turbulent velocity fluctuation (Reynolds stresses), term due to change in pressure and density and term is due to the shear stress tensor respectively. Since the source region is acoustically compact and the observer is in far field, Impermeable FWH is used. Farassat formulation 1A has been used as it is recommended for marine propellers (Farassat, 2007)., The pressure field is defined in the Farassat formulation as:

$$P'(\vec{x}, t) = P_T'(\vec{x}, t) + P_L'(\vec{x}, t) \qquad -(5)$$

where P' - acoustic pressure, $P_T'$and P_L' - acoustic pressure field resulting from thickness (monopole) and loading(dipole) The thickness and loading noise components are defined by Equation 6 & 7.

$$4\pi\, p_T'(x,t) = \int_{f=0} \left[\frac{\rho_0 . \dot{v}_n}{r(1-M_r)^2} + \frac{\rho_0 v_n \hat{r}_i \dot{M}_i}{r(1-M_r)^3}\right]_{ret} dS + \int_{f=0} \left[\frac{\rho_0 . c . v_n \,(M_r - M^2)}{r^2 (1-M_r)^3}\right]_{ret} dS \qquad \text{- (6)}$$

$$4\pi\, p_L'(x,t) = \int_{f=0} \left[\frac{\dot{p}.\cos\theta}{c.r(1-M_r)^2} + \frac{\hat{r}_i \dot{M}_i . p.\cos\theta}{c.r(1-M_r)^3}\right]_{ret} dS + \int_{f=0} \left[\frac{p.(\cos\theta - M_i n_i)}{r^2(1-M_r)^2} + \frac{(M_r - M^2).p.\cos\theta}{r^2.(1-M_r)^3}\right]_{ret} dS \qquad -(7)$$

The receivers are adjusted in six points above and behind the propeller at *x* and *z* axis (x/*D*, *z/D* =3, 4, 5, 6).The unsteady flow around the propeller models are simulated at J=0.833 with a moving reference frame at n= 10 rps. A time step of t= 10^{-4} sec is used. Hydroacoustic analysis has been carried out using FWH solver. SPL predicted are compared with the published numerical results of Seolet. al. (2002).

3. RESULTS AND ANALYSIS

A parametric study on the performance and hydroacoustic characteristics for DTMB 4119 propeller with and without tip rake (+/- 6^0 , +/-10^0) is carried out. The hydrodynamic performance for the DTMB 4119 propeller at J= 0.833 is found to be in good agreement with the published literature. The hydrodynamic characteristics of the DTMB propeller with tip rake (+/- 6^0 , +/-10^0) is also found. A comparison of the performance characteristics of the propeller post modification is tabulated (Table 2). There is a reduction in the overall efficiencies which can be observed except in case of propeller with tip rake -10^0 (forward rake). The primary reason for the reduction in efficiencies is the increase in the torque coefficient K_Q. The significant increase in Thrust Coefficient, K_T in case of -10^0 tip rake has resulted in keeping the overall efficiency same as the original propeller. Pressure distribution around the modified propeller is shown in Figure 4(a)(b). The pressure just above the blade tip and around pressure side is the maximum. The region between 0.7 to 0.9 R in the suction side of the blade has minimum pressure. The velocity distribution around the propeller is shown in Figure 5(a)(b).The blade tip produces high speed backward flow. At the same time a forward flow with lower axial velocity is also generated. These two reverse flows stimulate a vertical flow around the blade tip creating tip vortices. The variation in the pressures and velocities around the blade due to tip modification has resulted in the change in propeller performance. The flow around the blades creates the pressure fluctuations which act as noise sources. The SPL was estimated by acoustic analogy. The effect of tip rake angle will produce flow modifications and hence change in the SPL. The estimation has been carried out with density as 1000 kg/m3 and sound speed in water 1500 m/s. The reference pressure for SPL is 1.0 μPa. However, a reduction in SPL in the propeller with tip rake has been observed (Figure 6) at low frequencies (0 to 20 Hz) .The variation in SPL measured radially and axially are also plotted (Figure 7). The SPL is found to follow the inverse square law with distance. This means that if the distance doubles, the overall SPL reduces around 6–7 dB (Tian et al. 2014).It can be seen that both along shaft axis and propeller disk, the SPL reduces with increased distance from the source.

Table 2: Comparison of Propeller performance after introducing Tip rake

	Thrust Coeff, K_T	Torque Coeff, 10 K_Q	Open water efficiency ,μ_O
Zero rake	0.118	0.236	0.664
Tip rake - 6^0 (forward rake)	0.127	0.261	0.645
Tip rake +6^0 (backward rake)	0.116	0.243	0.634
Tip rake- 10^0 (forward rake)	0.138	0.275	0.664
Tip rake +10^0 (backward rake)	0.117	0.249	0.625

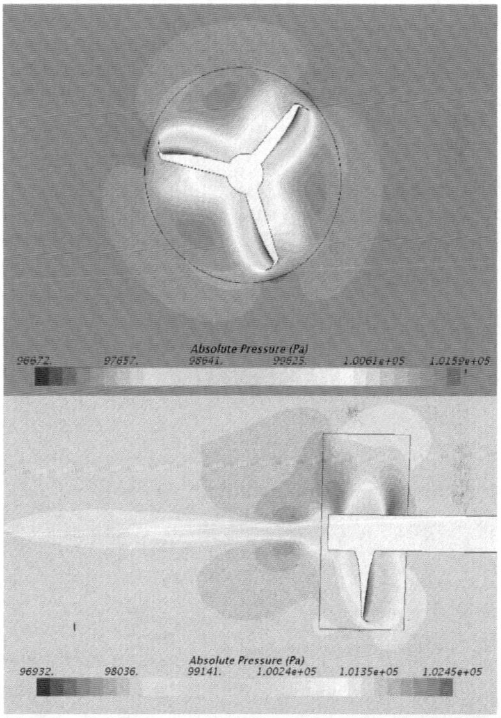

Figure 4: Pressure distribution around propeller blade at -6° Tip rake (a)Cross section (b) Axial

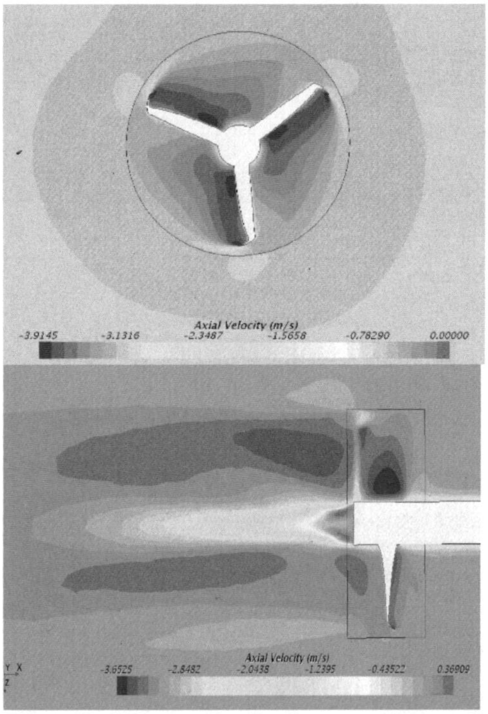

Figure 5: Velocity distribution around propeller blade at -6° Tip rake (a)Cross section (b) Axial

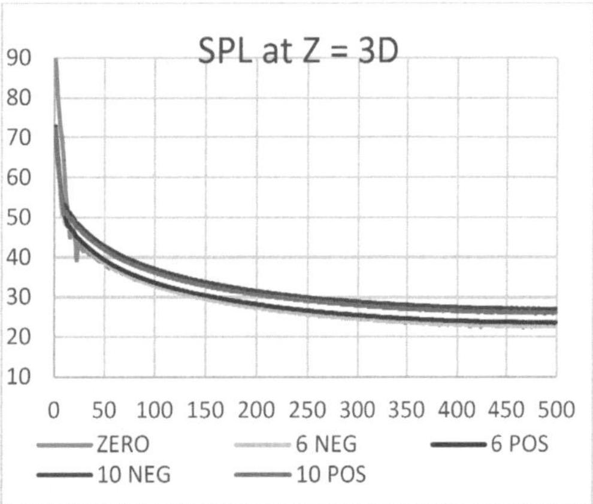

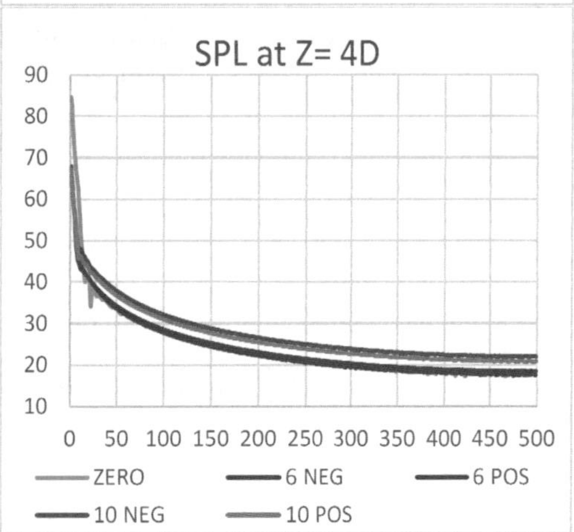

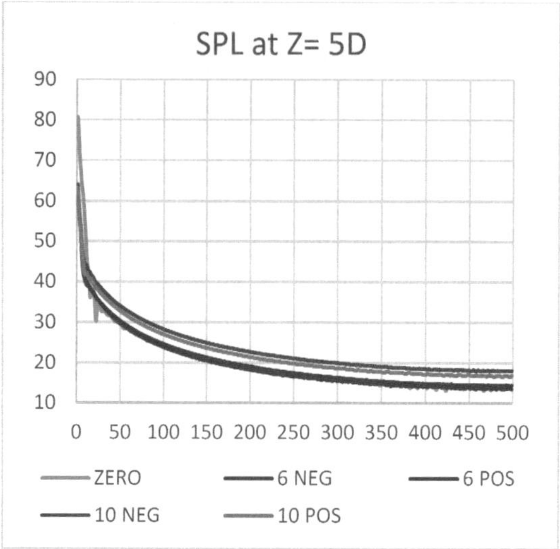

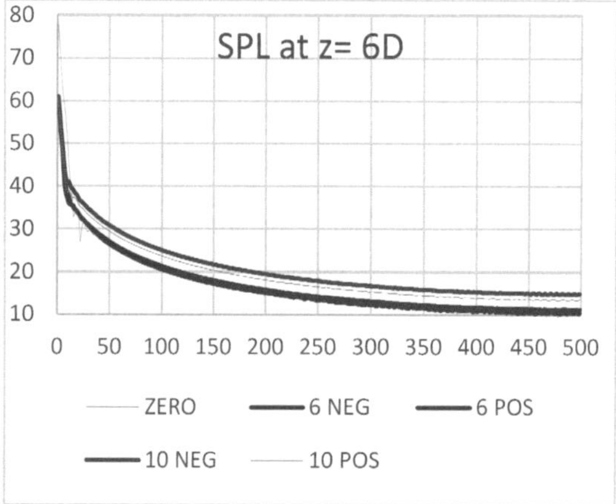

Figure6: Comparison of SPL for propeller with and without Tip rake (a) z= 3D (b) z= 4D (c) z= 5D (d) z= 6D

4. CONCLUSION

The hydrodynamic simulation was carried out on DTMB 4119 with and without tip rake angle using RANS solver and Realisable k-ε model for turbulence modelling and unsteady flow regime. The SPL was estimated using FWH equation at various points in axial and radial directions. Based on the obtained results, the following conclusions can be made:

(1) Hydrodynamic performance of marine propellers change with the introduction of tip rake.

(2) K_T and K_Q are modified with introduction of tip rake resulting in an overall reduction in propeller efficiency.

(3) The tip rake of -10° (forward rake) results in the same efficiency as that of the original propeller.

(4) The tip vortices are modified/shifted due to introduction of tip rake angle.

(5) Reduction in SPL of propeller with tip rake has also been observed at low frequencies(0 – 20 Hz). The reduction in SPL is about 10dB at 10 Hz.

(6) It can be seen that both along shaft axis and propeller disk, the SPL reduces with increased distance from the source.

The authors intend to work on studying the effect of tip modifications in different propellers both numerically and experimentally as the future work.

REFERENCES

[1] Carlton JS. (2012) Marine propeller and propulsion, 3rd ed. Elsevier publication Ltd.

[2] Richard, K. and Munjal, M. L. (1980) *Noise radiation from noncavitating marine propellers - A state of the art report.*

[3] Parchen, R. (2000) 'Noise production of ships propellers and waterjet installations at non-cavitating conditions', in *Proceedings of the 34th WEGEMT School 'Developments in the design of propulsors and propulsion systems'.*

[4] Cone J, Clarence D. 1962. The theory of induced lift and minimum induced drag of non-planar lifting systems, Technical Report, NASA-TR-R-139.

[5] Gennaro G, Gonzalez-Adalid J. 2012. Improving the propulsion efficiency by means of contracted and loaded tip (CLT) propellers. Athens.

[6] Sharma, S. D., Mani, K. and Arakeri, V. H. (1990) 'Cavitation noise studies on marine propellers', *Journal of Sound and Vibration*, 138(2), pp. 255–283. doi: 10.1016/0022-460X(90)90542-8.

[7] Kheradmand S, Rahrovi A,Mousavi B. 2014.Numerical simulation of tonal and broadband hydrodynamic noises of noncavitating underwater propeller.Polish Marit Res. 21(3):46–53.

[8] Li W, Yang C. 2009. Numerical simulation of flow around a podded propeller.Proceedings of the 9th International Offshore and Polar Engineering Conference; Jun 21–26; Osaka, Japan.

[9] LighthillM. 1952. On sound generated aerodynamically. I. General theory.Proceedings of the Royal Society of London; March 20; London.

[10] Hassan Ghassemi, Mohsen Gorji & Jalal Mohammadi (2018) Effect of tip rake angle on the hydrodynamic characteristics and sound pressure level around the marine propeller,Ships and Offshore Structures, 13:7, 759-768

[11] Gorji M, Ghassemi H, Mohammadi J. (2016) Determining the hydroacoustic characteristics of the ship propeller in uniform and nonuniform flow. Int J Eng Trans. A: Basics. 29(4):530–538.

[12] Pan YC, Zhang HX(2010) Numerical hydro-acoustic prediction of marine propeller noise. J Shanghai Jiaotong Univ. 15(6):707–712.

[13] Williams JE, Hawkings DL (1969) Sound generated by turbulence and surfaces in arbitrary motion. Phil Trans R Soc. 264(1151):321–342.

[14] Lidtke, A. K., Turnock, S. R. and Humphrey, V. F. (2015) 'Use of acoustic analogy for marine propeller noise characterisation', *Proceedings of the 4th Symposium on Marine Propulsors*, (June), pp. 231–239.

[15] Sezen, Dogrul and BalLloyd, T. P., Rijpkema, D. R. and van Wijngaarden, E. (2015) 'Marine propeller acoustic modelling: comparing CFD results with an acoustic analogy approach', in *Proceedings of the 4th International Symposium on Marine Propulsors.*

[16] ITTC (2014) 'ITTC – Recommended Procedures and Guidelines - Practical guidelines for ship CFD applications. 7.5-03-02-03 (Revision 01)', p. 19.

[17] Bagheri, M. R., Mehdigholi, H., *et al.* (2015) 'An experimental and numerical prediction of marine propeller noise under cavitating and non-cavitating conditions', *Brodogradnja*, 66(2), pp. 29–45.

[18] Caridi D. 2008. Industrial CFD simulation of aerodynamic noise [PhD thesis].Università degli Studi di Napoli Federico II.

[19] Farassat, F. (2007) *Derivation of Formulations 1 and 1A of Farassat, NASA.*

[20] Tian J, Yan H, Zhang Z, Yuan G, Rao Z, Hua H. 2014. LES-based numerical analysis of surface-pressure fluctuations and unsteady thrust of a marine propeller. In: Inter. noise. Melbourne, Australia.

[21] Tewari, A.K. and Vijayakumar,R.(2019) 'A Review of Methods for Hydro-acoustic analysis of Non-cavitating Marine Propellers' International Journal of Marine Engineering(accepted for publishing)

[22] Joe,D., Tewari,A.K. and Vijayakumar,R.(2018),'Numerical Study of Acoustic Characteristics of a Marine Propeller Due to Rake at Propeller Tip', International Conference of Computational and Experimental Marine Hydrodynamics MARHY 2018,Chennai, India.

9. AUTHORS BIOGRAPHY

Danio Joe is a PhD scholar at the Department of Ocean Engineering, Indian Institute of Technology Madras, India. His area of research is study of hydroacoustics of marine propeller. He has presented paper titled 'Numerical Study of Acoustic Characteristics of a Marine Propeller Due to Rake at Propeller Tip', at the International Conference of Computational and Experimental Marine Hydrodynamics MARHY 2018,Chennai, India in November 2018

Vijit Misra is a PhD scholar at the Department of Ocean Engineering, Indian Institute of Technology Madras, India. His area of research is hydro acoustics of marine propeller.

Dr R Vijayakumar is an Assistant Professor at the Department of Ocean Engineering, Indian Institute of Technology Madras, India. His area of interest includes warship and submarine design, autonomous underwater vehicles, seakeeping and manoeuvring of marine vehicles and underwater acoustics.

NUMERICAL ESTIMATION ABOUT PERFORMANCE OF CPP AND PODDED PROPULSOR ON MANY OFF-DESIGN CONDITIONS USING RANS APPROACH

K Himei, **K Okazaki** and **H Yamaguchi**, Nakashima Propeller Co., Ltd., Japan

SUMMARY

On the off-design condition of CPP or podded propulsor, the angle of attack against the propeller blade or the flow interaction between propeller and pod housing is quite different from the propeller design condition. The flow around them is very complex. Hence there is the difficulty for numerical estimating. In order to verify the applicability to wide range of the off-design condition, RANS simulations were executed. The calculation range for CPP was complete two quadrant condition each at several pitch setting. Regarding podded propulsor, pull type and push one having the common propeller and housing were targeted and the range of pod steering angle were set to all 360 degrees. Hydrodynamic performance as these calculated results were compared with published experimental results including the rotational moment of changing CPP's pitch angle or steering the pod housing. Then, it was found that there were good agreements except for the particular kind of the operating conditions.

NOMENCLATURE

$C_{0.7R}$	Blade chord length at 70% radius (m)
C_T	Thrust coefficient based on relative velocity, $C_T = T/\left(\frac{\pi}{8}\rho {V_2}^2 D^2\right)$
C_Q	Torque coefficient based on relative velocity, $C_Q = T/\left(\frac{\pi}{8}\rho {V_2}^2 D^3\right)$
C_{Qb}	Blade spindle torque coefficient based on relative velocity, $C_{Qb} = Q_b/\left(\frac{\pi}{8}\rho {V_2}^2 D^3\right)$
D	Propeller Diameter (m)
J	Propeller advance coefficient, $J = V_a / (nD)$
K_{MZ}	Pod steering moment coefficient, $K_{MZ} = M_Z / (\rho n^2 D^5)$
K_T	Thrust coefficient, $K_T = T / (\rho n^2 D^4)$
$K_{T(POD)}$	Thrust coefficient of pod housing, $K_{T(POD)} = (T\cos(\psi) - T_X) / (\rho n^2 D^4)$
$K_{TX,TY,TZ}$	X, Y, and Z components of force coefficient, $K_{TX,TY,TZ} = T_{X,Y,Z} / (\rho n^2 D^4)$
K_Q	Torque coefficient, $K_Q = Q / (\rho n^2 D^5)$
n	Rotating number of propeller (s^{-1})
$M_{X,Y,Z}$	Moment around X, Y and Z-axis (N m)
$P_{0.7R}$	Blade pitch as 70% radius (m)
Q	Propeller torque (N m)
Q_b	Blade spindle torque (N m)
$R_n(k)$	Kempf's Reynolds number, $R_n(k) = C_{0.7R}\sqrt{{V_a}^2 + (0.7\pi nD)^2}/\nu$
T	Propeller thrust (N)
$T_{X,Y,Z}$	X, Y and Z components of force (N)
V_a	Inflow velocity of water (m/s)
V_r	Relative velocity at 70% radius (m/s), $V_r = \sqrt{{V_a}^2 + (0.7\pi nD)^2}$
y^+	Dimensionless wall distance
β	Hydrodynamic pitch angle of the inflow, $\beta = \tan^{-1}\{V_a/(0.7\pi nD)\}$
η_o	Propeller open efficiency, $\eta_o = J \cdot K_T / (2\pi K_Q)$
ν	Kinematic viscosity coefficient of water (m^2/s)
ρ	Mass density of water ($kg\ m^{-3}$)
ψ	Steering angle of Pod propulsor

1. INTRODUCTION

When designing Controllable Pitch Propeller (hereinafter called CPP) or Podded propulsor, the possible performances, especially about the moment for rotating the blade pitch angle or pod housing on any off-design conditions, must be considered properly [1]. Because the estimated result of that spindle moment on early design phase directly leads to the selection of the hydraulic control system capacity. During proceeding in parallel with the hydrodynamic design approach and the structural design approach as a mechanical device, this planned capacity has a high impact on physical size and total cost of propulsor. Unfortunately, if an excessive hydraulic system is planned on early design phase, too robust and too large CPP hub or pod housing finally lead to decrease the propulsion efficiency too. In addition, the off-design performance of these propulsors is important not only for vessel's propulsive performance but also for manoeuvring and dynamic positioning performance of vessels.

However, it is difficult to numerical estimate their hydrodynamic performance values under wide range with high accuracy because the flow field around CPP's blade or pod housing is so complicated. Therefore, in many cases, these estimations have relied on actual equipment data as past design results or on design chart based on model tests. Some CPP's representative charts as open water characteristics were published by Yazaki et al [2] as AU-CP 4 series and Chu et al [3] as 3-bladed JD-CPP series. Although it doesn't cover all the operating conditions of CPPs since there are the lack of systematic information. And as the application of FPP chart to CPPs, Wageningen B-Series [4] is also often used. However the effect of characteristic CPP blade form and relatively large size of CPP hub compared with FPP cannot be considered. The experimental data of the blade spindle moment [5][6][7] are also not enough to designing new contemporary CPPs. Therefore, since there were the demands of wide range systematic information including with the off-design condition, MARIN was developed Wageningen C-Series and carried out the complete two quadrant open water tests [8] in recent years.

While on the other hand, about podded propulsor, very limited systematic information can be found. The effects of geometry variations of pod body on the hydrodynamic performance were researched by Islam et al [9]. However, to the knowledge of the authors, there are no published experimental studies that the principal particulars or geometrical shape of propeller blades was changed systematically to research the influence to open water performances.

There are only a few published studies on computational approaches [10][11][12]. Moreover, the calculation ranges are limited and do not cover with all operating conditions. The reason seems to be the difficulty of numerical calculation, too much number of analysis required for evaluation, the lack of experimental data due to the particularity of measurement.

In this paper, a CPP and a couple of podded propulsor (pull type and push one, same propeller part) with experimental results were analysed by RANS simulations. The ranges of calculation condition were set very widely. Hydrodynamic characteristics and the blade spindle torque or steering moment of pod housing are compared with measured values.

2. ANALYSIS CONDITIONS FOR CPP

RANS simulations were executed using the commercial CFD software, "SC/Tetra Ver.14", which is based on a finite volume method with an unstructured grid.

2.1 PROPELLER

In this paper, results of RANS simulation for propeller model 7166R of Wageningen C-Series [8] are compared with experimental measurements. The photograph and principal particulars of model 7166R are shown in Figure 1 and Table 1. Among many models in Wageningen C-Series, in this paper, the model 7166R was chosen as for being the closest principal particulars to CPP adopted for high speed vessels such as ferries and Ro/Ro ships. In the simulations, the geometric shape of the blade completely matched the model propeller with the inclusion of the consideration about the blade tip rounding and the thickness of the trailing edge and so on. On the other hand, the gap between the hub and the blade palm was ignored and always smoothed in simulations.

2.2 SETUP CONDITIONS FOR CPP

Table 2 shows the main setting conditions for RANS simulations for CPP. The rotating number of propeller n and the inflow velocity V_a were set so that Kempf's Reynolds number $R_n(k)$ matched the experiments at each conditions of hydrodynamic pitch angle of the inflow β. $R_n(k)$ varies with β, and its value is between about 3.8×10^5 and 7.7×10^5. In this range of $R_n(k)$, it is expected that both laminar flow area and turbulent flow area exist on propeller blade surface. LKE k-k$_L$- ω turbulent model [13] was selected to simulate the transitional flow. The transition flow analysis using LKE k-k$_L$- ω model in SC/Tetra was shown by Hasuike et al [14]. While, the analyses applied SST k- ω turbulent model [15] also were carried out. The application of this model was to consider the effect of "Shear Stress Transport" in turbulent flow. Especially in the case which the absolute value of β was large, it was purpose to improve the analysis accuracy of the flow separation from blade at the adverse gradient region. Analysis results using these two models are compared in the next chapter. The prismatic cells near propeller surface were arranged for resolving the boundary layer. The first layer thickness of the prismatic cells was set to a non-dimensional wall distance for a wall-bounded flow (y^+ in short) equal to 1.0.

Figure 1 Photograph of model 7166R of Wageningen C-Series [8].

Table 1 Principal particulars of model 7166R.

Number of blades	4
Diameter, D	212.38 mm
Pitch at 70% radius, $P_{0.7R}$	212.38 mm
Pitch ratio at 70% radius, $P_{0.7R}/D$	1.000
Expanded blade area ratio	0.700
Hub diameter	58.00 mm
Hub-prop diameter ratio	0.2731

Table 2 Setup conditions for CPP.

Type of analysis	Steady-state
Kempf's Reynolds number	Same as experiments
Turbulent models	LKE k-k$_L$- ω model and SST k- ω model
y^+	1.0

2.3 DOMAIN AND MESH ARRENGEMENT

In RANS simulations, the large stationary domain and the inner rotational domain including propeller were set as shown in Figure 2. These two cylindrical domains were connected by sliding mesh technique. And the size of them was adopted large enough to ignore the effects on thrust or torque generated by propeller. For the same reason, the length of propeller hub was modelled long enough so that a uniform flow comes into the blade root.

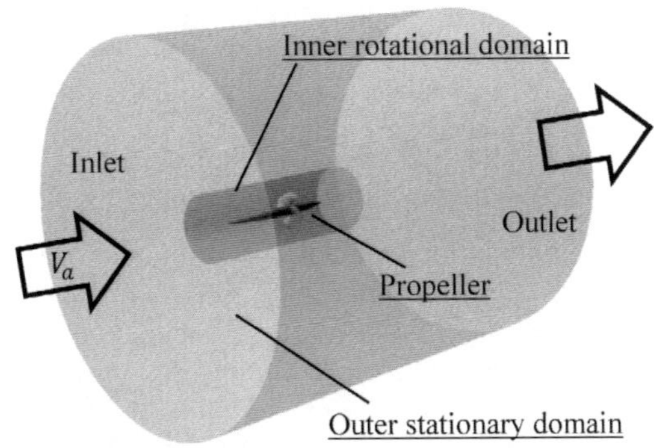

Figure 2 Computational domain for CPP.

The numerical mesh was an unstructured grid, and basic cells were tetrahedral. As above, the prismatic cells were applied to near the blade surface. These were enough fine to almost ignore the mesh size dependency as shown in Figure 3. The total number of meshes was about 33 million at any pitch angle setting.

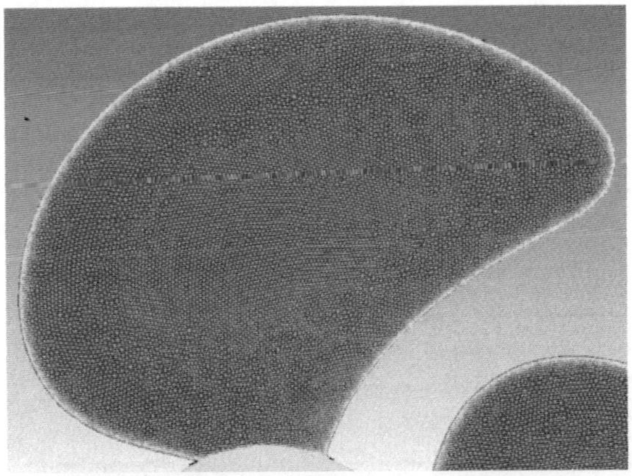

Figure 3 Numerical meshes near CPP blades.

3. OPEN WATER PERFORMANCE OF CPP ON OFF-DESIGN CONDITIONS

3.1 PROPELLER OPEN CHARACTERISTICS AT DESIGN PITCH

First, to verify the set turbulent models on the normal operating condition including propeller design point, the comparison of Propeller Open Characteristics at design pitch setting and in the range of $0 < J < 1.0$ as shown in Figure 4. $J = 1.0$ corresponds to $\beta = 24.5$ degrees. The results of LKE k-k$_L$-ω model agree with experimental results quite well. However, there are the underestimation of K_T and the overestimation of K_Q when using SST k-ω model. Figure 5 shows calculated limiting streamlines on blade. While LKE k-k$_L$-ω model could simulate the laminar-turbulent transition on the model scale, the result of SST k-ω model was almost fully turbulent flow. It is found that the analysis using the model that can treat the transition flow like LKE k-k$_L$-ω model was suitable when evaluating the geometric shape of propeller blade and its propeller open efficiency in detail on the normal operating condition and on the model scale.

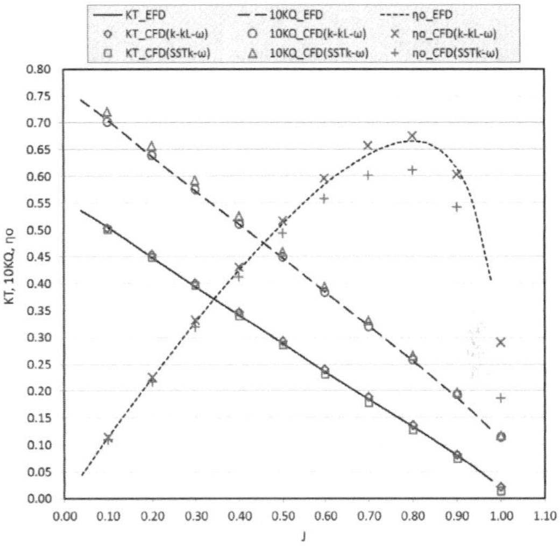

Figure 4 Propeller Open Characteristics at design pitch setting.

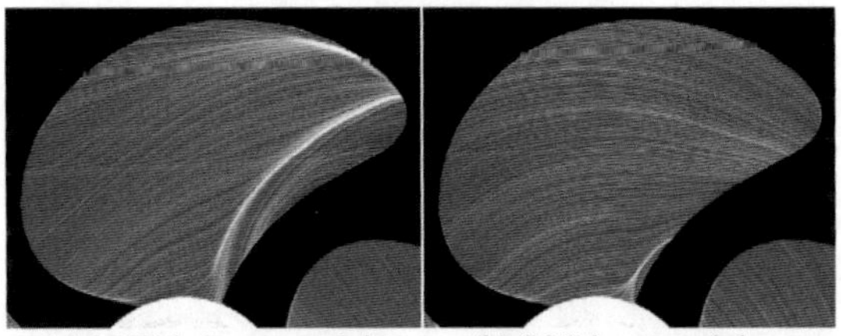

(a) LKE k-k$_L$-ω model (b) SST k-ω model

Figure 5 Limiting streamlines on blade.

3.2 PROPELLER PERFORMANCE IN TWO QUADRANT CONDITION AT CHANGING PITCH SETTING

RANS simulations in complete two quadrant condition, $-90 \leq \beta \leq 90$ degrees, were executed each at four different pitch setting including the design pitch angle and minus pitch setting. Their results are shown in Figure 6. For both LKE k-k$_L$-ω model and SST k-ω model, Trends of calculated results approximately matched that of the experiments. Needless to say, RANS simulation, unlike applying the design charts based on past model propeller, can completely reflect the any blade shape designed each opportunity. Especially when considering the application to the plan for various operating capacities on early design phase, it is found that RANS simulation would be sufficiently valuable as estimating the hydrodynamic propeller performances including blade spindle torque on CPP off-design condition.

Comparing both turbulent models, SST k-ω model agrees with better in range where the absolute value of β is large. This suggests the importance of the treatment for separation flow on off-design conditions where the angle of attack to blade is unusual.

4. ANALYSIS CONDITIONS FOR PODDED PROPULSOR

4.1 PROPULSOR

The 5 bladed podded propulsors of both push propeller type and pull propeller type that Heinke carried out the model tests at 360 degrees steering angle [16] were analysis object. The propeller part, model CP1375, has the same geometric shape in both type, and its principal particulars are shown in Table 3. The outline view of overall pod propulsors and stationary coordinate system are shown in Figure 7. X-axis is longitudinal, Y-axis is transverse and Z-axis is vertical direction. The positive direction of the steering angle ψ is counter-clockwise viewed from the top.

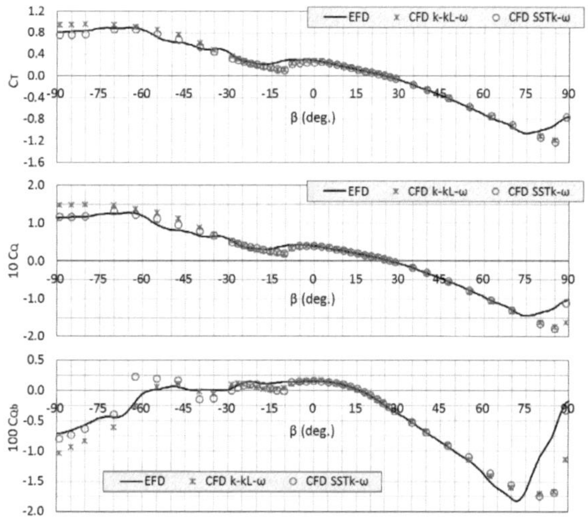

(a) pitch setting : $P_{0.7R}/D = 1.0$ (design pitch angle)

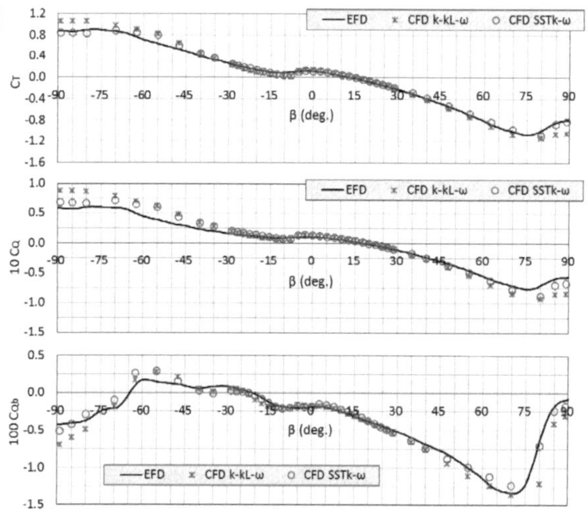

(b) pitch setting : $P_{0.7R}/D = 0.5$

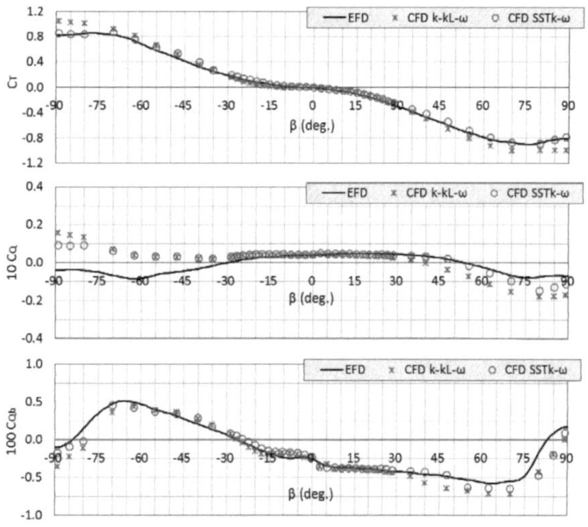

(c) pitch setting : $P_{0.7R}/D = 0.0$

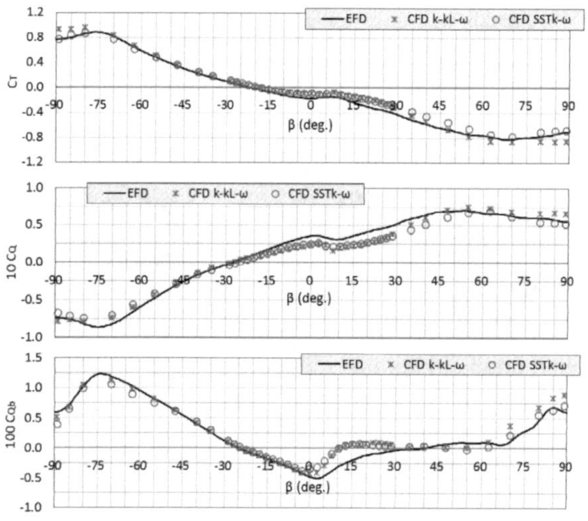

(d) pitch setting : $P_{0.7R}/D = -0.7$

Figure 6 Hydrodynamic characteristics at each pitch angle setting.

Table 3 Principal particulars of model CP1375.

Number of blades	5
Diameter	220.00 mm
Pitch at 70% radius	244.86 mm
Pitch ratio at 70% radius	1.1130
Expanded blade area ratio	0.8109
Hub diameter	56.50 mm
Hub-prop diameter ratio	0.2568
Propeller design point	0.847 (pull type) 0.850 (Push type)

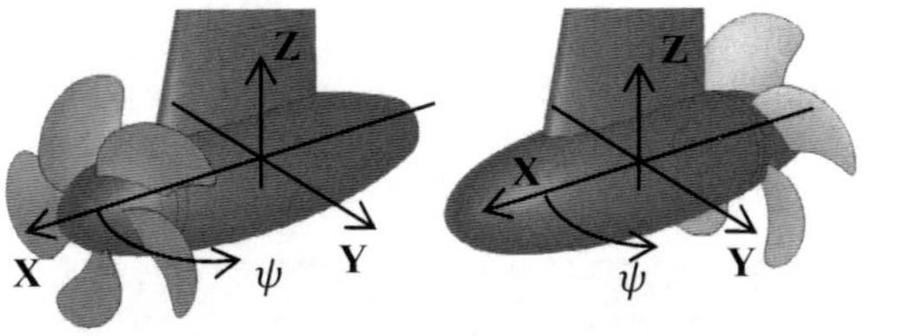

Pull type Push type

Figure 7 Pod outline view and stationary coordinate system.

4.2 SETUP CONDITIONS FOR PODDED PROPULSOR

Table 4 shows the main setting conditions for RANS simulations for podded propulsor. In unsteady analysis, it is necessary to sufficiently develop the flow created by the pod propulsor before the hydrodynamic propeller performance evaluation. Since the pod body is long in the inflow direction, it is important to balance the total number of cycles for flow development and practical analysis cost. In this paper, Realizable k- ε model [17] which is one of the high Reynolds number models was applied, and in order to avoid the too many total numbers of prismatic cells in the boundary layer, y^+ was set about 50. Realizable k- ε model imposes restrictions on physical realizability by correction for eddy-viscosity so that the turbulent energy k does not become a negative value, and therefore it is said that this model are effective for analysis of the swirling flow. Time step per cycle was defined as rotational step of propeller, and set about 1.5 degrees step in initial stage of analysis, after then shortened to 0.5 degrees step in evaluation stage. Total number of cycle was determined so that the flowing distance was about 5 times the pod body length or more. This corresponds to about 16 rotations of the propeller at the CP1375 propeller design point J = about 0.85.

The inner rotational domain including only propeller part was set, and connected with the outer large stationary domain by using sliding mesh as same as CPP analysis. The numerical meshes, in addition to making them near the blades and pod housing fine, were arranged while paying attention to the spatial resolution which becomes the upstream and downstream region of the pod housing and propeller blades depending on steering angle. Figure 8 shows the part of numerical meshes near propeller blades and pod body. The total number of meshes was about 30 million for any steering angle analysis.

Table 4 Setup conditions for Podded propulsor.

Type of analysis	Unsteady-state
Kempf's Reynolds number	6×10^5
Turbulent models	Realizable k-ε model
y^+	About 50
Time step	0.5 deg. rotational step per cycle

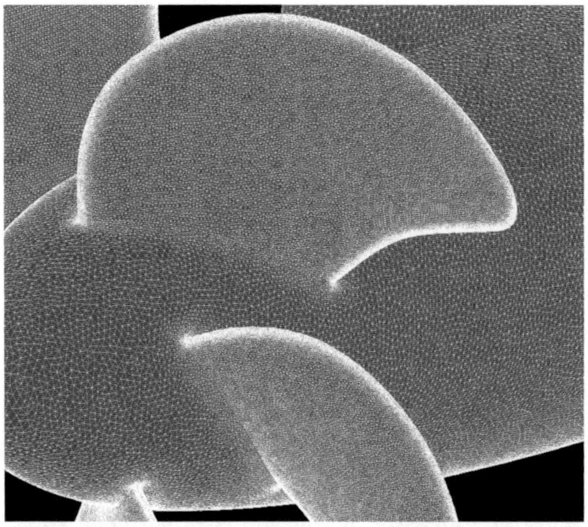

Figure 8 Numerical meshes near propeller blade and pod body.

5. OPEN WATER PERFORMANCE OF POD PROPULSOR AT ANY STEERING CONDITION

5.1 AT ZERO STEERING ANGLE

Comparison between calculated and measured propeller open characteristics at zero steering angle is shown in Figure 9. In case of pull type, calculated K_T was slightly smaller than measured values, Calculated $K_{T(POD)}$ corresponds approximately to measured values, and calculated K_Q overestimated in whole. One of reason for K_Q difference seems to overestimate of the frictional component of propeller torque by fully turbulence flow condition caused by using Realizable k- ε model for high Reynolds number. On the other hand, in case of push type, K_T and K_Q difference between calculated and measured value were the opposite trend to pull type. As a result, η_o was higher than measurement despite the overestimation of the frictional torque component. It seems to be caused by analysis error on the wake flow from the pod housing located in front of the propeller part.

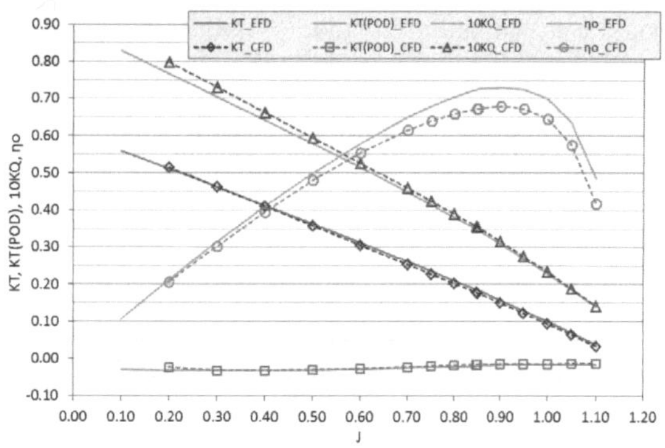

(a) Pull type

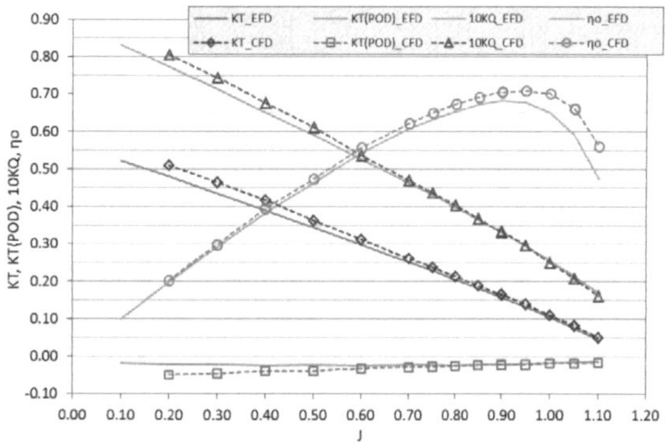

(b) Push type

Figure 9 Comparison of propeller open characteristics.

5.2 AT ANY DEGREES STEERING ANGLE

Figure 10 shows the comparison of the each force and moment coefficients at the propeller design point for the steering angles ranging from 0 to 360 degrees steering angle. Roughly overviewing, RANS simulations could catch the distribution trend of each force and moment associated with changing steering angle. In the range of $-90 \leq \psi \leq 90$ degrees, calculated pod steering moment coefficient K_{MZ} of pull type are somewhat underestimated but RANS simulations of both type have good agreements with experimental results. This steering angular range covers most of the practical operation conditions.

However, in the range of $90 < \psi < 270$ degrees, the agreements are not so good and the distributions of calculation results vary widely. Some additional simulations were carried out with more cycles. However, no more consistent results were obtained. As ψ changes, the angle of attack of propeller blade becomes extremely large. For instance when ψ = 180 degree, propeller operating condition is similar to reverse sign of propeller design point, that is J = -0.85. Therefore, on above range of ψ, the flow near the whole blade is very complex associated with the flow separation. It is assumed that there is room for improvement on the setup condition and mesh resolution.

The value of K_{MZ} of pull type is maximum when ψ=75 degrees, and about 5 time as compared with push type as shown in Figure 10. Figure 11 shows comparison of the contour maps of surface pressure on the inlet side at ψ =75 degrees. White curves are streamline near the propeller part. It is found that accelerated swirling wake flow greatly interferes with the pod housing behind propeller in case of pull type, therefore there is the significant difference in surface pressure of pod housing. The positive pressure, where on inlet side of pod surface between steering center and the propeller part, makes K_{MZ} large. In addition, since the sizes of these pod housings are relatively large with respect to propeller diameter, it is susceptible to propeller wake flow. As a result the ratio of K_{MZ} maximum value is large.

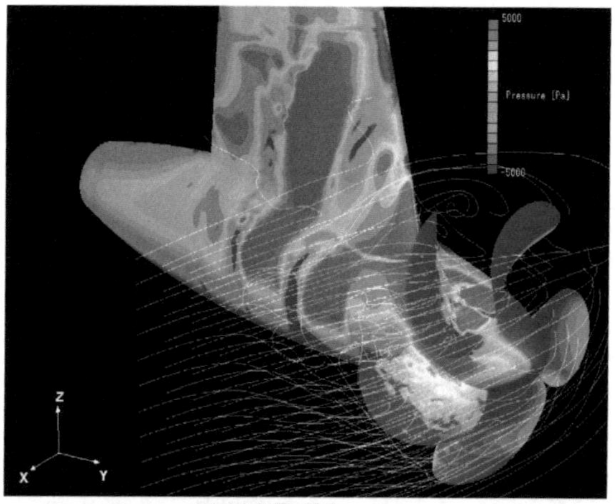

(a) Pull type

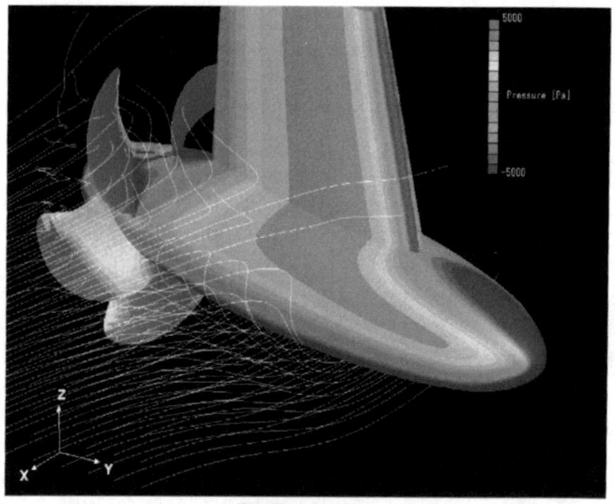

(b) Push type

Figure 11 Contour maps of surface pressure at ψ =75 degrees.

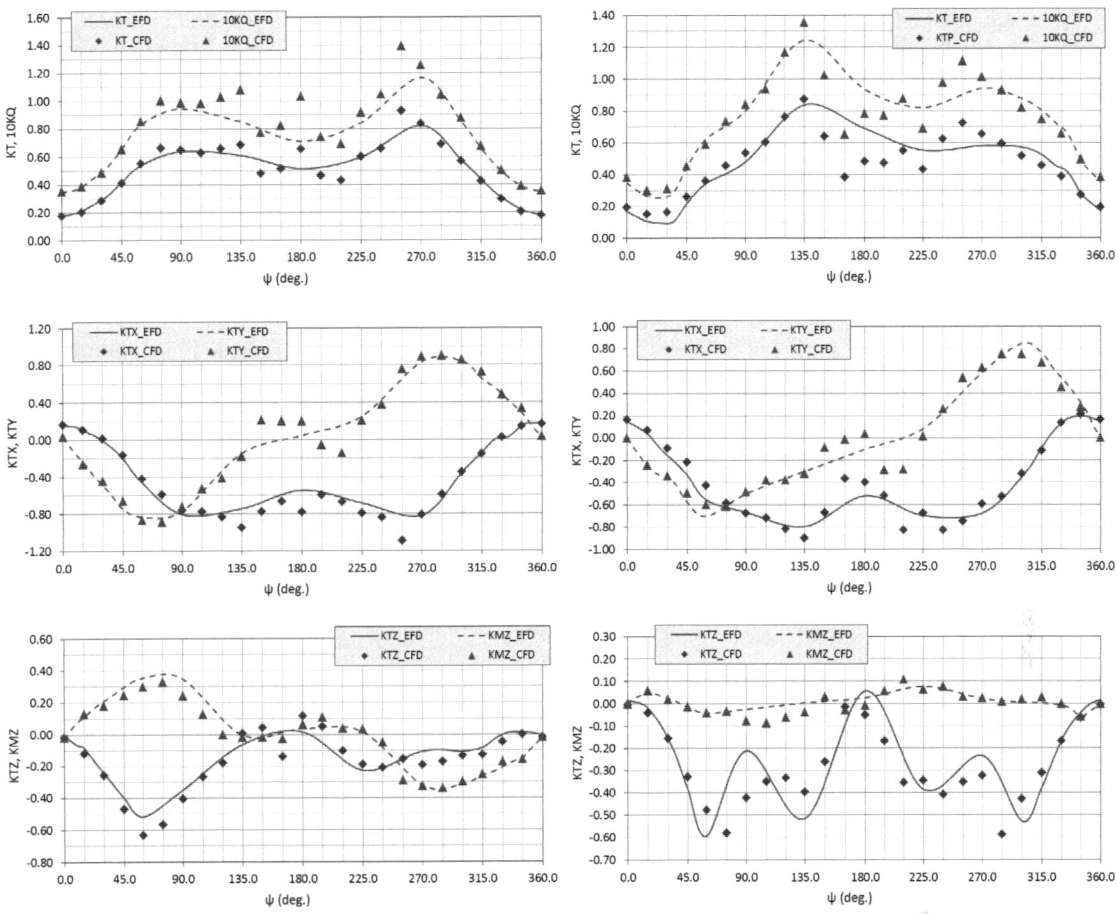

(a) Pull type **(b) Push type**

Figure 10 Comparison of force and moment coefficients.

7. CONCLUSIONS

RANS simulations were executed to analyze CPP and podded propulsor on various off-design conditions. Main conclusions are as follows;

1) RANS simulation for a CPP in complete two quadrant condition could approximately catch the trend of experiments at any pitch angle setting. It is considered to be usable for overall performance estimations at the initial design phase.

2) It is important that the turbulence model suitable for the CPP operating condition is selected. The calculation results were influenced whether the laminar-turbulent transition or the separation flow could be analysing well.

3) RANS simulation for podded propulsor also had good agreements with measured value within 90 degrees steering angle. Most of the podded propulsor's actual operation condition corresponds to this steering angular range. Rough trends throughout all steering angle could be obtained, however in angle range over 90 degrees, the dispersion of calculated results was large and there was no high-accuracy matching to experiments.

8. ACKNOWLEDGEMENTS

The authors would like to acknowledge the kindly cooperation's of Mr. Hans- Jürgen Heinke and Mr. Ulf Barkmann for the provisions of the geometric data of podded propulsor and the detail information's of model test setting.

9. REFERENCES

1. Pronk, C., 'BLADE SPINDLE TORQUE AND OFF-DESIGN BEHAVIOUR OF CONTROLLABLE PITCH PROPELLERS', *Doctoral thesis of Technical University Delft*, 1980.

2. Yazaki, A., Kuramochi, E. and Osaki, T., 'Design Diagrams of Four-bladed Controllable Pitch Propellers', *Journal of the Society of Naval Architects of Japan, Vol. 112*, 1962.

3. Chu, C., Chan, Z. L., She, Y. S. and Yuan, V. Z., 'The 3-bladed JD-CPP series', *Proceedings of the 4th LIPS Propeller Symposium*, 1979.

4. van Lammeren W.P.A., van Manen J.D. and Oosterveld M.W.C., 'The Wageningen B-screw Series', *Transactions of SNAME, Vol. 77*, 1969.

5. Denny, S. B. and Nelka, J. J., 'BLADE SPINDLE MOMENT ON A FIVE-BLADED CONTROLLABLE-PITCH PROPELLER', *DTMB Report No.3729*, 1972.

6. Boswell, R. J., Nelka, J. J. and Kader, R. D., 'EXPERIMENTAL SPINDLE TORQUE AND OPEN-WATER PERFORMANCE OF TWO SKEWED CONTROLLABLE-PITCH PROPELLERS', *DTMB Report No.4753*, 1975.

7. Ito, M., Yamasaki, S., Oku, M., Koizuka, H., Tamashima, M. and Ogura, M., 'An Experimental Study of Flow Around CPP Blade (3rd Report): Measurement of CPP Blade Spindle Torque', *Journal of the Kansai Society of Naval Architects, Vol.192*, 1984.

8. Dang, J., Brouwer, J., Bosman, R. and Pouw, C., 'Quasi-Steady Two-Quadrant Open Water Tests for the Wageningen Propeller C- and D-Series', *Proceedings of the 29th symposium on naval hydrodynamics*, 2012.

9. Islam, M. F., 'performance study of podded propulsors with varied geometry and azimuthing condition', *Doctoral thesis of Memorial University of Newfoundland*, 2009.

10. Funeno, I., Pouw, C. and Bosman, R., 'Measurements and Computations for Blade Spindle Torque of Controllable Pitch Propellers in Open Water', *Proceedings of the 3rd International Symposium on Marine Propulsors, SMP'13*, 2013.

11. Shamsi, R. and Ghassemi, H., 'Numerical investigation of yaw angle effects on propulsive characteristics of podded propulsors', *International Journal of Naval Architecture and Ocean Engineering, Vol.5*, 2013.

12. Liu, A., Dang, J., Xie, Q. and Hu, J., 'The Effect of Sheet Cavitation on the Blade Spindle Torque of a Controllable Pitch Propeller', *Proceedings of the 4th International Symposium on Marine Propulsors, SMP'15*, 2015.

13. Walters, D. K. and Cokljat, D., 'A Three-Equation Eddy-Viscosity Model for Reynolds-Averaged Navier–Stokes Simulations of Transitional Flow', *ASME, Journal of Fluids Engineering, Vol.130*, 2008.

14. Hasuike, N., Okazaki, M., Okazaki A. and Fujiyama, K., 'Scale effects of marine propellers in POT and self propulsion test conditions', *Proceedings of the 5th International Symposium on Marine Propulsors, smp'17*, 2017.

15. Menter, F. R., 'STwo-equation eddy viscosity turbulence models for engineering applications', *AIAA Journal, Vol. 32, no.8*, 1994.

16. Heinke, H. J., 'INVESTIGATIONS ABOUT THE FORCES AND MOMENTS AT PODDED DRIVES', *Proceedings of T-POD, First International Conference on Technological Advances in Podded Propulsion*, 2004.

17. Shih, T. H., Liou, W. W., Shabbir, A., Yang, Z. and Zhu, J., 'A New k- ε Eddy Viscosity Model for High Reynolds Number Turbulent Flows Model Development and Validation', *Computers Fluids, Vol. 24, No. 3*, 1995.

9. AUTHORS BIOGRAPHY

Kohei Himei is manager of propulsion performance section at Nakashima Propeller and researcher on various marine propulsor. He has been studying on hydrodynamics performance of CPP, thruster and podded propulsor in addition to FPP since seven years ago. As a result, it contributed to the development of high-performance five blades CPP. One of his study themes is also a partially submerged surface piercing propeller.

Kosuke Okazaki and **Hirotaka Yamaguchi** belong to Propulsion System Design Dept. at Nakashima Propeller and are in charge of CPP blade design and CFD operations.

RIM DRIVEN PROPELLERS: OPTIMIZATION BASED DESIGN APPROACH USING RANS CALCULATIONS

S Gaggero, University of Genoa, IT

SUMMARY

RIM driven propellers represent an unconventional, but underrated, propulsive solution which hydrodynamic design is still not obvious. In the last years, most of the attention has been devoted to the efficient coupling with electric motors, since only recently the development of permanent magnets allowed for the successful embedding of the driver directly inside the surrounding duct. From the hydrodynamic point of view, however, analysis and, in particular, design strategies are not yet ripe. In the light of the development of advanced design approaches for unconventional geometries, we propose a Simulation-Based Design Optimization tool based on RANS analyses of parametrically described geometries as a part of an automatic, multi-objective optimization loop. The SBDO is used to design a RIM driven propeller with an accelerating type duct, with improved performance simultaneously in terms both of efficiency and cavitation inception at different (design and quasi-bollard pull) functioning conditions.

1. INTRODUCTION

RIM driven thrusters represent one of the most innovative propulsive concepts of the last years, even if the idea of driving a propeller via its blade tips dates back to the work of Saunders [1]. Preliminary applications can be recognized in some USSR ships of half the past century [2,3] but only recently, thanks to the availability of materials and innovative manufacturing technologies, such as brushless, permanent magnet based, electric motors [4], the functional implementation of this kind of propulsors was possible. The interest toward this type of propulsors was mainly driven by the increased attention that the underwater radiated noise from ships has received over the last decade. Improved naval system, as well as more comfortable cruise vessels but, primarily, the possible impact on marine wildlife, focused the attention to propulsive solutions obviously capable of maintain high level of efficiency (to comply with recent EEDI and EEOI IMO indexes but at an even lower level of radiated noise, which is considered definitively a pollutant like any other emissions to air and water. But designing for an increase of the cavitation-free speed normally leads to a reduction in propulsive efficiency and, then, to a considerable increase in fuel burn and emissions, which is completely against current ambitions to reduce emissions of GHG, NOx, SOx, and particulate matter.

The use of non-conventional propulsive solutions like contra-rotating and tip loaded propellers [5], together with Energy Saving Devices like PBCF [6, 7] or Mewis Ducts [8] and modern optimization-based design approaches attempt mainly to comply with the propulsive efficiency issues and in some cases, with the combination of different solutions or thanks to tailored design approaches, a simultaneous reduction of pressure pulses and underwater radiated noise was possible [9, 10].

RIM driven thrusters, which from the hydrodynamic point of view are very similar to ducted propellers, could represent a further step towards efficient and silent propulsive solutions. A RIM driven thrust, in fact, is a propeller inside a nozzle (or a tunnel) driven from the tip of the blades by an electric motor, which permanent magnetic rotor is embedded in a "band" on the blade tips while the stator is mounted on the duct. Both rotor and stator are canned separately: a rotating seal is then not needed and being the electric motor fully surrounded by seawater also cooling issues are significantly reduced. Contrary to usual ducted configurations, there is also no gap between the propeller tip and the inner surface of the nozzle. This means that local energy losses can be minimized by the almost vanishing of both the characteristic tip vortex form blade tip and the tip leakage vortex generated in the gap of ducted configurations [11, 12]. The possible increase of efficiency (which however could be affected by the additional torque associated to the frictional forces acting on the tip ring), then, simultaneously supports the reduction of the radiated noise and the resulting vibrations normally associated to tip vortexes.

The advantages of RIM configurations, moreover, are not limited to noise and vibrations reduction: a compact and flexible installation (thanks to the reduction of auxiliary systems), larger speed range and increased maneuverability performance associated to the torque response of the permanent magnet electric motor are some of the positive aspects associated to RIM propulsive configurations [13]. Without struts and shaft, in addition, the inflow to the propeller is more uniform and, in the case of tunnel thrusters, pressure head losses can be minimized. However, also due to the scarce literature concerning this type of propulsive solution, RIM driven thrusters rise a lot of hydrodynamic challenges in both design and performance analysis compared to largely studied conventional propellers. Most of the studies, indeed, concerned mainly with the design and analyses of the electric motor system rather than on the design of the most efficient, cavitating free, tip driven propeller blade.

Main hydrodynamic issues regard the blade loading distribution and the "rim effect" associated with the flow (and the resulting friction losses) between the rotor and the stator of the electric motor.

Obviously, the blade loading of a RIM driven propeller is significantly different from that of a conventional configuration. Without any clearance between the inner surface of the nozzle and the blade tip, the radial circulation which is zero at the tip for conventional propellers, has a finite value and, under certain conditions of pitch, chord and camber distributions, could reach its maximum [14]. Zero circulation, on the contrary, characterizes the blade root, increasing the risk of stronger root vortexes. The opportunity to increase (or to maintain sufficiently high) the load in the outer part of the blade is beneficial from the point of view of propulsive efficiency. With the gap between the blade and the inner surface of the nozzle sealed, moreover, the risk of tip vortexes is significantly reduced (and the leakage avoided) also in case of high loads. The hydrodynamic design of the blade, however, has to consider that the higher load could lead to excessive values of suction pressure and a higher risk of (bubble) cavitation, which from the noise point of view is as harmful as the cavitating tip vortex.

The first attempts to design RIM driven propellers having in mind these opportunities, which easily can turn into new design constraints, were those of Sparenberg [15] and Mishkevich [16]. Under the assumption of finding the optimal radial distribution, they used modified lightly and moderately loaded lifting line design approaches to characterize both the blade and the nozzle loading distributions. Yankolev et al. [3] proposed an optimization based design approach using, however, for a RIM driven tunnel thruster operating close to bollard pull condition, an even simplified hydrodynamic model consisting in 2D calculations for a lattice of hydrofoils. The merit function, under the constraint of a given thrust coefficient, was a weighted mean between efficiency and minimum pressure coefficient over the blades to delay cavitation as much as possible. The pitch distribution was the only design parameter of the proposed design approach. Recently also RANS calculations were employed to predict the hydrodynamic performance of RIM propellers, but only a few different geometries and functioning conditions were addressed [14]. With the exception of the analyses by Cao et al. [14], there is no evidence of detailed studies aimed at characterizing the modified flow on the outer sections of the blade and the related tip vortex phenomena, especially for a design point of view, which are a key feature of RIM propellers since most of the advantages in terms of noise and pressure pulses rely exactly on the avoidance of these phenomena possible by the RIM arrangement. RANS analyses were employed, instead, to quantify the "rim" effect, which is the inhibitor of the efficiency gain achieved with tip loading and mitigation of tip vortexes. The presence of the radial and axial gaps between the rotor and the stator determines torque losses, which have been preliminarily included in design methods using Taylor-Couette flow analogies between coaxial cylinders. Most of the RANS analyses of RIM propellers available in the literature [14, 16], similarly, make use of these approximations since the explicit simulation of the flow regimes in the annulus region which would result prohibitively expensive was used only to tailor simplified correlations.

Having in mind the potentialities and the possible issues of RIM driven propellers, in this work a hydrodynamic design approach based on optimization strategies similar to those developed for conventional, non-conventional, ducted propellers and energy saving devices [5, 6, 7], is proposed for this kind of propulsors. The opportunity of such a design approach is dictated by the outcomes of the few works dealing with RIM driven propellers. Little, in fact, was published on the design philosophy of this kind of propulsors. Only general design procedure involving, in particular, the electric/electromagnetic aspects, were developed for symmetric (port/starboard performance) propulsors somehow derived from the KA propeller series, operating in straight tunnels at bollard pull. The few design approaches available in the literature, moreover, are based on potential approaches which may suffer from too crude approximations especially when the design space is widened to explore unusual configurations for efficiency and cavitation improvements. Instead, the use of RANS calculations for design purposes proposed in this work allows to exactly address the peculiarities of RIM driven thrusters that would be exploited for performance maximization. It is believed that, exactly as demonstrated in the case of other propellers type, an optimization based design approach could provide the necessary flexibility for a multidisciplinary problem, allowing to simultaneously consider the possible interactions of the well-established electromagnetic design issues with the hydrodynamic and the structural demands of well-performing RIM blades. Even if in this work only the hydrodynamic aspects are considered throughout the design process, the potentialities of such advanced design approach capable of dealing with the multiple objectives (maximization of efficiency, as usual, with minimization of the risk of any type of cavitation) needed to fulfill current requisites will be demonstrated.

2. RIM DRIVEN PROPELLER

As described in the introduction, a RIM driven propulsor is a propeller inside a nozzle, driven, rather than by the hub, from the tip of the blades, exactly through the "rim" (a kind of movable ring on the internal surface of the nozzle) to which all the blade are joined. A schematic representation of a RIM driven propeller is given in Figure 1. RIM blades are attached to the rim (in orange), which rotates with the blades and contribute to the total absorbed torque. Its width obviously changes in order to accommodate the blades.

$$K_{T\,total} = K_{T\,RIM\,blades} + K_{T\,Nozzle} + K_{T\,RIM}$$
$$K_{Q\,total} = K_{Q\,RIM\,blades} + K_{Q\,RIM} \qquad (1)$$

The arrangement of Figure 1 resembles that adopted for the design using the Simulation Based Design Optimization framework proposed in this paper. The design by optimization is carried out for a four bladed-propulsor (Z=4) operating in a MARIN 19A accelerating type nozzle, which obviously contributes to the total thrust (Eq. 1).

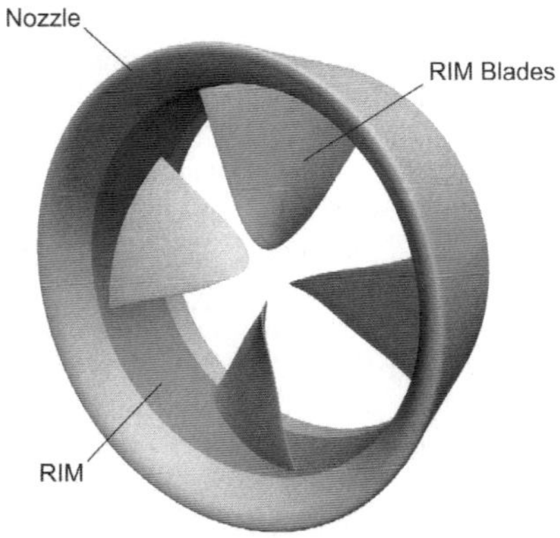

Figure 1: Example of mesh arrangement for RIM driven propeller analysis.

2. SBDO APPROACHES FOR DESIGN PURPOSES

In the last decades, the design of conventional and non-conventional marine propellers has significantly changed. Traditional design tools were developed mainly for the maximization of efficiency but (and sometimes especially) propeller side effects, like pressure pulses and radiated noise, are becoming the most important constraints in order to comply with comfort requirements and even stricter environmental regulations. Avoiding (or at least controlling) cavitation in unsteady functioning, designing highly loaded efficient propellers for multiple functioning points and minimizing, at the same time, the risk of excessive vibrations onboard, are some of the main issues a designer has to deal with and with respect to which traditional design tools show all their limitations.

Simultaneously, the propeller design process, similarly to what happened for many complex engineering systems, has been substantially modified also by the availability of new simulations tools which can comply with (at least some of) the recent design requirements. The increased robustness and accuracy of numerical algorithms together with the exponential development of hardware resources foster the application of simulation methodologies capable of reducing (or even replacing) the experimental load by providing most of the data necessary for the assessment of the additional design criteria and objectives, already in an early design stage, to address the growth of technical and commercial requirements.

Simulation-Based Design (SBD) frameworks have, consequently, become a standard also for propellers design in order to provide an environment for efficient connections among tools and approaches for multi-disciplinary analyses using advanced simulations, aimed at reproducing the phenomena of interest with a significant degree of accuracy. Obviously, the highest fidelity at the lowest possible cost is essential but not trivial. Accurate design analysis methods are computationally expensive but necessary to provide the flexibility and the validity needed in case of novel concepts and unconventional configurations which, instead, represent the application limits of computationally efficient, low-order, approaches. Establishing the performance of a design, or the relative merits of trade-off alternatives, when a decision about a design path has to be faced, is crucial in the constant search of improvements imposed by increased market competition. Then, larger design spaces and more alternatives have to be explored and compared, each time by using higher fidelity simulations and complex geometrical representations, to allow for potentially bigger improvements. In this context, simulation-based design frameworks easily turn into simulation-based design optimization frameworks (SBDO), where simulation tools, optimization algorithms and parametric descriptions of the geometry are fully and automatically integrated to deal with large numbers of decision parameters and stringent and mutually conflicting design objectives and constraints.

Dealing with high-dimensional and large design spaces is necessary because the expensive exploration of a wider design space considering unconventional configurations far from "usual design rules" and accepted guidelines is, very often, the simplest way for improving performance. But using medium-fidelity computation approach inherently narrows the exploration of the design space in terms of possible functioning conditions under investigation, especially in the case of unconventional configurations, since it is mandatory to avoid as much as possible designs optima falsified by the insufficient accuracy and the inherent limitations of the medium fidelity flow solver. This is the reason which recommended the use of RANS calculations (Section 2.2) for this specific design of a RIM driven propeller in place of the more computationally efficient BEM analyses, previously adopted [5, 6, 9] in similar SBDO.

Instead, the parametric description of the RIM blade can be handled using, as usual [5, 9] B-Spline curves (Section 2.1) approximating the classical design table. Also in the case of RIM driven propellers, the design table in a certain way can be considered as the inherent parametric description of the blade. In addition, the flexibility and the geometric variability provided by the B-Splines representation is almost unrivaled by other parametric descriptions: the relatively low number of control points needed for the parametrization of the reference propeller provided by this type of curves allowed for not excessively large design spaces maintaining, at the same time, a design variability consistent with the requirements of fairness generally expected for propeller blade shapes.

The entire process, finally, from pre-processing of geometries to post-processing of RANS results and monitoring of constraints and objectives, has been arranged within the modeFRONTIER environment [18] using a genetic type optimization algorithm (Section 2.3).

2.1 PARAMETRIC DESCRIPTION OF THE RIM DRIVEN PROPELLER GEOMETRY

The parametric description of the RIM driven blade follows the well-established guidelines already exploited for BEM based SBDM design strategy. B-Spline curves are used to approximate the most relevant distributions of the design table (i.e. chord, pitch, and maximum sectional camber), since for the design activity a non-skewed, zero raked blade was considered. Control points of B-Spline polygons turn into the free parameter of the optimization, leading to an 18 dimension design space.

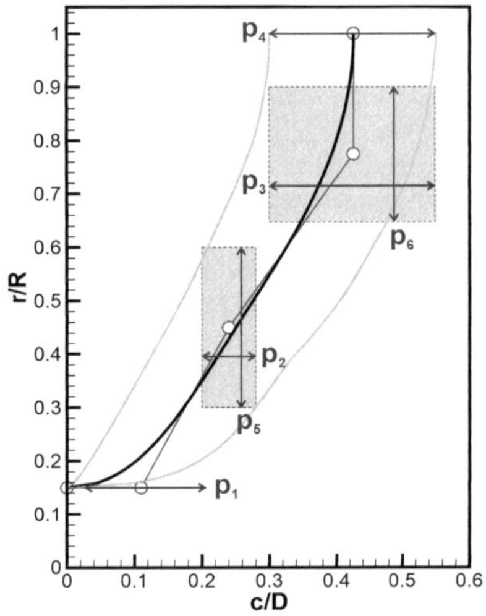

Figure 2: Parametric description of the chord distribution. Range of variations of parameters and envelope of all possible distributions.

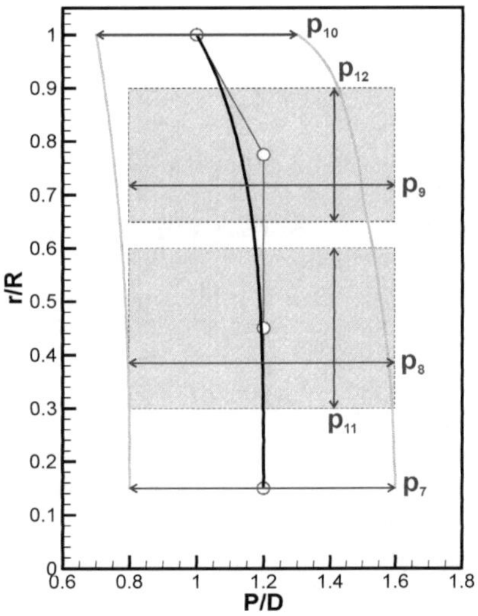

Figure 3: Parametric description of the pitch distribution. Range of variations of parameters and envelope of all possible distributions.

As shown in Figures 2, 3 and 4, modification of the chord, as well as of the pitch, are allowed thanks to 6 parameters, while 5 free parameters handle the maximum sectional camber. The eighteenth parameter consists of a scaling factor for the leading edge radius. Sectional profiles belong to NACA66 family for what regards the thickness distribution. Camber line has a conventional NACA a=0.8 shape. Range of variations, highlighted in the figures, were selected to comply with the geometrical constraints (i.e. at the blade/RIM junction no change of the radial position of the parameter is possible, as well as at the blade root) and to define geometries strictly included in the simplest periodic computational domain consisting in an angular sector of aperture equal to $2\pi/Z$ (Section 2.2). The result of the parametric description is a stereolithographic (.STL) representation of the RIM blade, ready for the meshing process (Figure 5), together with the edge features that are useful for local cells refinements.

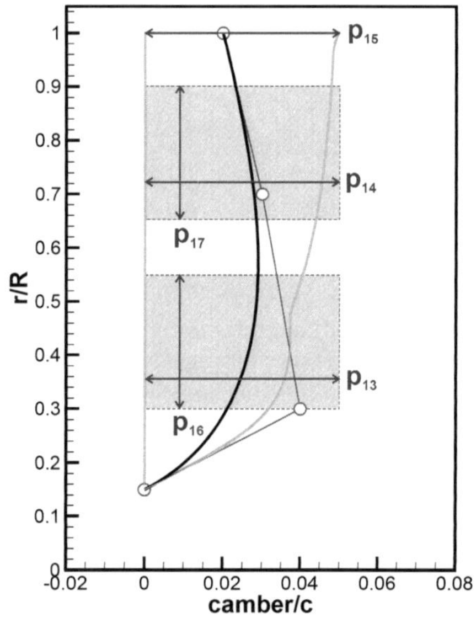

Figure 4: Parametric description of the maximum section camber distribution. Range of variations of parameters and envelope of all possible distributions.

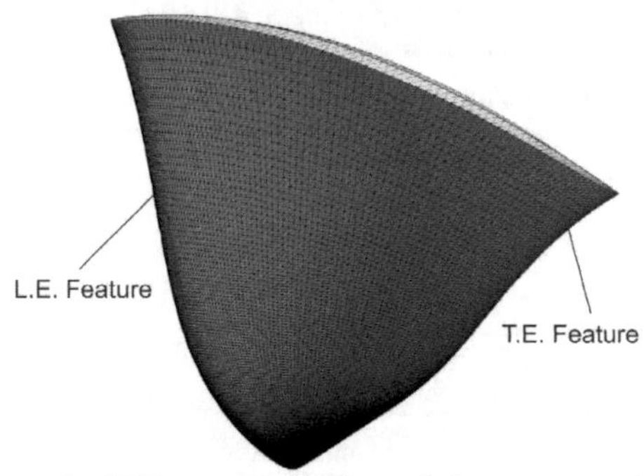

Figure 5: STL representation of the parametrically modeled RIM blade.

2.2 NUMERICAL TOOLS FOR RIM DRIVEN PROPELLER ANALYSIS

Under the hypotheses of incompressible and steady flow, the time-averaged continuity and momentum equations (RANS) can be written in the differential form of Eq. (2):

$$\begin{cases} \frac{\partial V_i}{\partial x_i} = 0 \\ \rho V_j \frac{\partial V_i}{\partial x_j} = -\frac{\partial p}{\partial x_i} + \frac{\partial}{\partial x_j}\left[\mu\left(\frac{\partial V_i}{\partial x_j} + \frac{\partial V_j}{\partial x_i}\right)\right] + \frac{\partial \tau_{ij}}{\partial x_j} \end{cases} \quad (2)$$

where $x_{i=1,2,3}$ is the spatial Cartesian coordinate and $V_{i=1,2,3}$ the Cartesian components of the velocity vector in the body-fixed reference frame. The fluid static pressure and density are, respectively, p and ρ, while μ is the fluid viscosity and τ_{ij} are the Reynold stresses resulting from the averaging process of the momentum equations, which are treated on the basis of the Boussinesq hypothesis, then proportional to the trace-less mean strain rate tensor times the eddy viscosity μ_T.

In the present analyses, the OpenFOAM package [19] was used to solve Eq. (2) on unstructured, hexa-dominant, finite volume meshes realized using the open source package cf-Mesh [20]. By exploiting the periodicity of the propeller geometry, the computational domain consists only in a blade passage handled with periodic boundary conditions. The computational domain has a cross area at the propeller plane one hundred times higher than the area of the propeller disk (D_{domain}/D_{prop} =10), the inlet is placed four propeller diameters in front of the propeller disk, the outlet four propeller diameters aft, and uniform inflow is assumed as the initial condition.

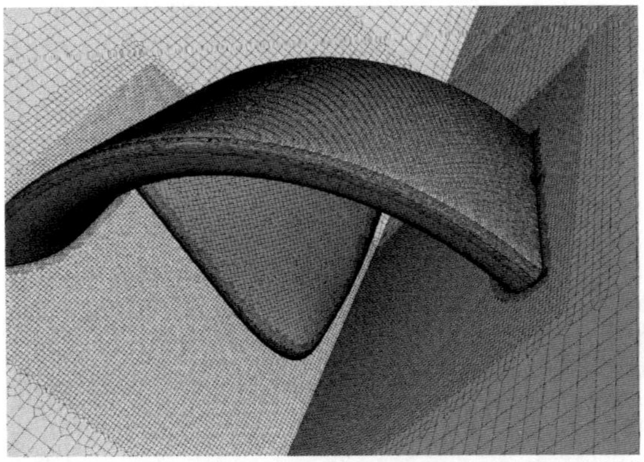

Figure 6: Example of mesh arrangement for RIM driven propeller analysis.

The meshing parameters, selected on the basis of preliminary analyses and previous calculations [21] are summarized in Table 1. Local refinements are applied at the leading and trailing edges of the blades (where feature edges have been identified during the parametric description of the blade) as well as of the accelerating nozzle surrounding it, while volumetric refinements allow for a better resolution on the propeller wake. This choice of parameters determines an average non-dimensional wall distance of about 25, handled using the wall functions from the SST k-ω turbulence model, for a total number of cells around 1.3 – 1.5 Million, depending on the shape of the RIM blade under investigation.
An example of the resulting mesh is shown in Figure 6.

Table 1: Meshing Parameters.

	Value
Avg. cell size (unrefined)	13% of D
Avg. cell size (around the propeller)	1.5% of D
Avg. cell size (on propeller wake)	4.5% of D
Avg. cell size (on propeller streamtube)	8.5% of D
Avg. cell size (on RIM and Duct)	1.5% of D
Min. cell size (on RIM and Duct LE & TE)	0.15% of D
Number of Prism layers	4
Prism layers stretching factor	1.2

The solution is computed using the steady solver simple FOAM. A moving reference frame, with the inclusion of the relative source terms in the RANS equations, allows for steady-state calculations. The solution is computed in a segregated form, by using the "consistent" version of the SIMPLE algorithm to handle the velocity-pressure coupling, and by applying under-relaxation (0.9 for velocity, 0.6 for turbulent quantities) to achieve a steady state solution. Second, upwind biased (for momentum) and first (for turbulent quantities) order accurate schemes are employed to discretize equations. Gradients are approximated with the Gauss quadrature and the multi-dimensional limiter. Numerical schemes for Laplacian and surface normal gradients are selected in order to account for non-orthogonal meshes.

2.3 OPTIMIZATION STRATEGY, OBJECTIVES AND CONSTRAINTS

The core of the SBDO approach is the optimization algorithm, which is in charge of driving the design process towards the optimal configuration by satisfying the design constraints and by selecting a geometry which is a trade-off between all the opposite objectives required for a specific application.
In the specific case under investigation, the design of the RIM driven propeller operating in an accelerating duct was carried out by requiring a propulsor configuration able to deliver, at the design advance coefficient (J) equal to 0.8, a total thrust coefficient (blades plus duct, $K_{T\,total}$) equal to 0.30 with the maximum possible efficiency and with the highest possible cavitation inception velocity. At the same time, the propulsor is required to deliver the highest possible thrust close to the bollard pull condition, which is a functioning point that has been monitored by a dedicated simulation at J equal to 0.3 to avoid the possible issues of boundary conditions reliability and convergence when the inlet velocity is excessively slow.
The resulting workflow is that of Figure 7. The process has been subdivided into seven steps, two of which determines if calculations can continue until convergence or not. After meshing (step 1-3), a preliminary analysis of the performance of the devised geometry using stable, first

order, numerical schemes is carried out for 400 initial iterations (step 4).

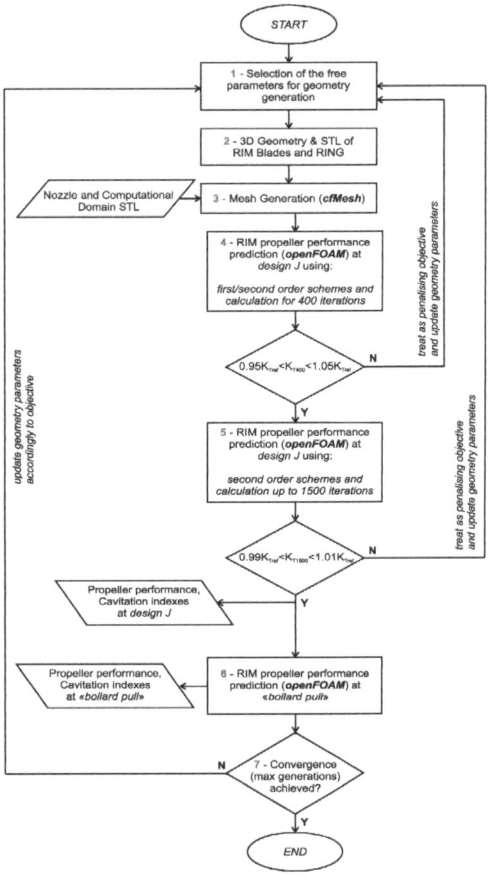

Figure 7: Example of mesh arrangement for RIM driven propeller analysis.

Only if propeller performance satisfy a sort of "relaxed" design constraints (total delivered thrust within 5% of the design target), calculations can continue for the last 1100 iterations (step 5) necessary to achieve converged results using second-order accurate numerical schemes. At this point, if the tight constraint on thrust (±1% of the design target) is satisfied, efficiency and cavitation inceptions are collected and the quasi bollard pull condition (step 6) is addressed. On the contrary, the geometry is discarded and its results used as penalizing objectives.

Since during the design process only non-cavitating calculations are considered to have a computationally efficient process, cavitation is monitored through the simplest inception criteria, then by the non-cavitating pressure coefficient. Collecting the minimum of the pressure coefficient in principle would be sufficient since cavitation starts when and where the pressure is below the vapor pressure. However, in order to have a better representation of the phenomenon (i.e. depending on the position of the minimum, be able to discern between leading edge sheet cavitation or bubble midchord cavitation), the RIM blade has been subdivided into seven different zones (Figure 8) for each of which the values of pressure (pressure coefficient) have been collected and monitored throughout the design process. Sheet cavitation is monitored on the suction and on the pressure side, close to tip (tip zone, r/R>0.9) and in correspondence of midspan (r/R between 0.5 and 0.9) for a total of four monitored values of pressure coefficient ($C_{PN1back}$, $C_{PN3back}$, $C_{PN6face}$, $C_{PN7face}$). Midchord cavitation, only on the suction side, is verified at the tip and at midchord ($C_{PN2back}$, $C_{PN4\ back}$) while $C_{PN5back}$ takes into account the minimum of the pressure coefficient close to the blade "root".

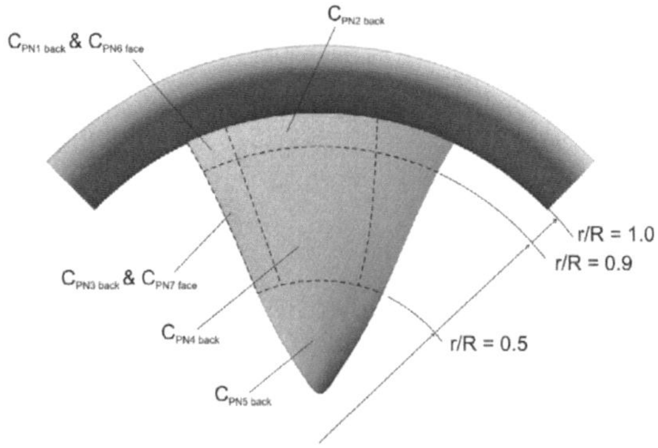

Figure 8: Blade zones subdivision for cavitation inception monitoring.

The design process, in the end, is configured as a constrained multi-objective optimization process, with one constraint (delivered thrust at the design advance), nine objectives (maximization of efficiency at the design advance, minimization of the seven suction pressure peaks on the blade at the design advance, maximization of the thrust at quasi bollard pull condition) and eighteen design parameters, which has been addressed using a multi-objective genetic algorithm on an initial Design of Experiments of 180 configurations and 20 generations, for a total of 3600 evaluated geometries.

3. RESULTS

A collection of results (filtered by the ranges of the objectives) from the design activity is shown in the parallel diagram of Figure 9. Orange lines represent (the performance of) unfeasible candidates which did not satisfy the relaxed threshold on thrust (step 4) and that consequently were discarded. Grey lines, instead, stand for feasible designs fulfilling the design constraints.
Some tendencies of the design process can be observed at first simply by looking at how designs are distributed in terms of design objectives. While for what regards cavitation inception on the suction side results are quite clustered towards higher value of pressure (i.e. lower risk of cavitation), filtered geometries show scattered data in terms of suction peaks on the face side together with values of suction significantly higher. In the design process, in a certain sense, it was easier to identify configurations able to minimize the risk of suction side cavitation while small variations of the design parameters,

well tolerated in terms of suction side phenomena, lead to a significantly increased risk of pressure side cavitation.

Among these feasible geometries, 5 candidates (ID_2165, ID_2527, ID_2958, ID_4919 and ID_4966) have been selected for additional detailed analyses. The choice has been carried out, for ID_2165 and ID_4966, in order to postpone as much as possible any cavitation inception, regardless of its location. ID_2165, in this respect, represents the best overall configuration, able to avoid cavitation on both suction and pressure side, up to a cavitation index of 2.55. By relaxing this threshold and assuming a design cavitation index of 2.82, among the few geometries able to avoid any cavitating phenomena at this functioning point, ID_4966 is the one with the highest efficiency, which is 2.4% higher (54.8% for ID_2165, 57.4% for ID_4966) than the efficiency provided by the ID_2165 configuration. The last three geometries (ID_2527, ID_2958 and ID_4919) were selected, instead, by relaxing only the inception of face-related cavitation phenomena. Fixing the inception threshold on the face (blade tip region) equal to 3.32, ID_2527 appears as a geometry able to realize intermediate values of efficiency (55.5%) with better margins for what regards back side phenomena: on the suction side, indeed, the occurrence of cavitation is postponed to a cavitation index of 2.40 at the leading edge of the tip region while midchord, bubble type, cavitation occurs only if the cavitation index falls below 2.30. ID_2527 delivers, moreover, the highest thrust at quasi-bollard pull condition. Increasing the risk of face cavitation to 3.50 allows to identify ID_2958, which is the worst configurations among those identified in terms of cavitation inception (on the suction side cavitation is anticipated to 2.66 for leading edge sheet cavitation, to 2.80 for what regards midchord bubbles) but that is able to deliver the required thrust with the highest efficiency (58.5%). Finally, ID_4919 is the geometry which provides the best performance on the suction side (inception postponed up to cavitation indexes of 2.30 at the leading edge, 2.15 at midchord) at the cost, however, of significantly anticipated phenomena on the pressure side (inception at 3.66 for sheet cavitation at the tip). In Figures 10 and 11, it is possible to appreciate the comparisons among these configurations and easily identify the zones (and their extensions) where the different cavitating phenomena take place. A reference geometry of a conventional ducted propeller operating at a slightly higher advance coefficient but delivering the same total thrust is moreover added to the comparison.

From the comparison, it is possible to highlight some, sometimes obvious, tendencies of the design process, which however are handled automatically in the selection process without the need of iterative and manual adjustments to get the results under the design constraints. Higher margins against cavitation are achieved using propellers with higher expanded area ratio which, in turns, reduces the overall propulsive efficiency. By comparing the overall best RIM design, ID_2165, with the reference ducted propeller, an evident reduction of the risk of cavitation can be appreciated on both suction and pressure side. The reference propeller has a large zone over the tip where the pressure is significantly lower (inception index of 3.36) than in the case of the RIM configuration (inception index of 2.55). This improvement is achieved with larger chords at the tip that, without the gap between the propeller and the nozzle, can better sustain higher loads without the risk of tip and leakage vortexes. Also, the pressure side of the RIM driven configuration shows a more favorable pressure distribution since for the reference ducted propeller it is possible to appreciate significant pressure reduction at the leading edge (inception index anticipated to 4.75) versus values lower than 2 in the case of the ID_2165.

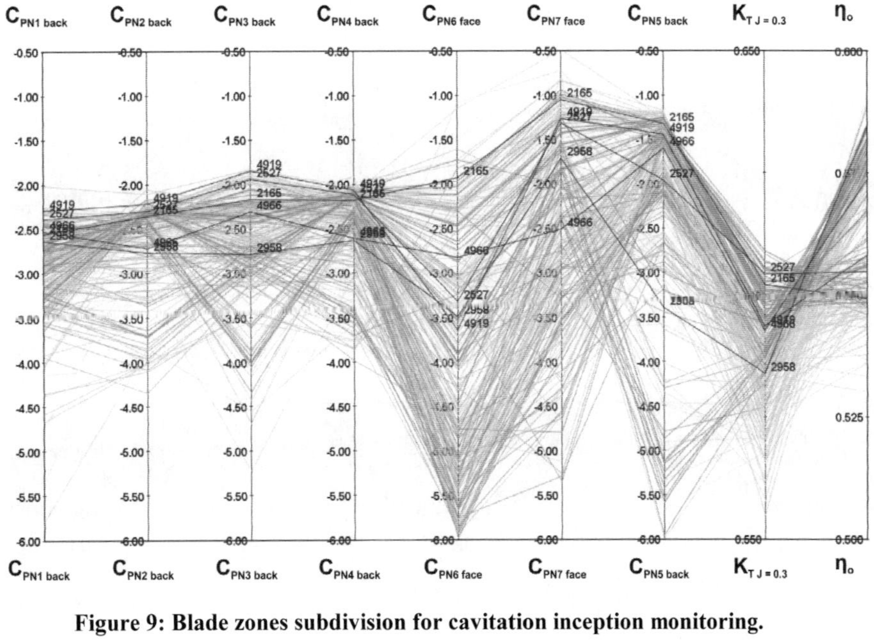

Figure 9: Blade zones subdivision for cavitation inception monitoring.

Efficiency reduction is the cost of these improvements since the RIM driven configuration has an efficiency significantly lower (54.8%) than that of the conventional ducted propeller (59.1%, considering also the hub). Two are the causes of this efficiency reduction. The first is the higher expanded area ratio, which increases the wetted blade area and then the frictional forces. This effect, however, is not the most relevant one, since if the RIM blades only are considered for computing the efficiency (and then neglecting the contribution of the absorbed torque of the rotating "rim" to which the blades are joined), the propulsive efficiency of RIM configurations increases to 59.5% for ID_2165 and reaches values higher than 61% for the most efficient solutions (ID_2958 and ID_4966).

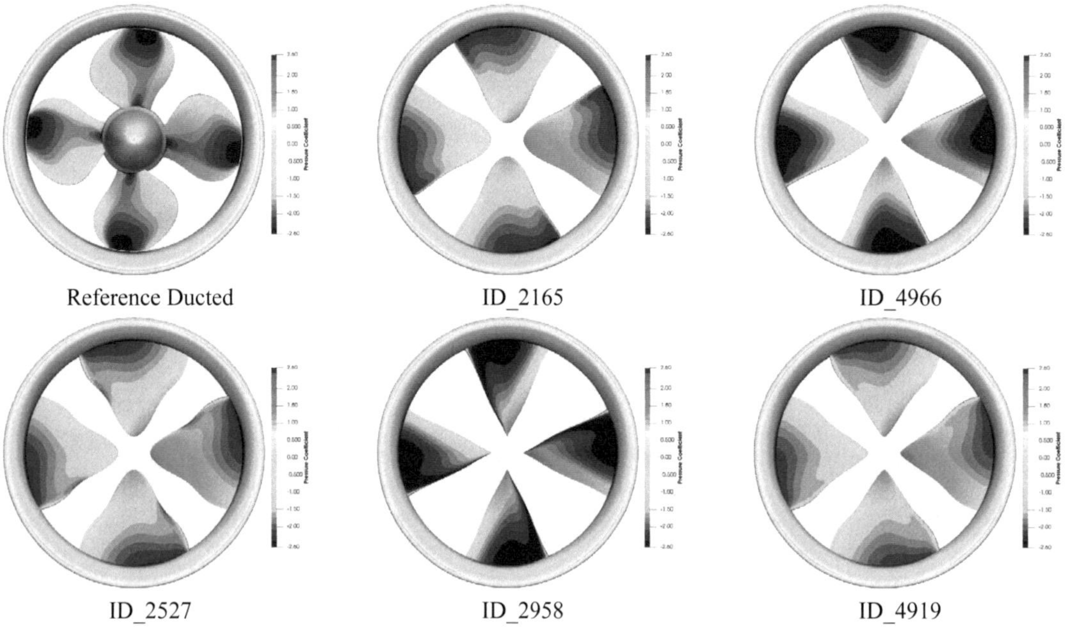

Figure 10: Pressure coefficient distributions on the selected RIM driven propellers. Suction side.

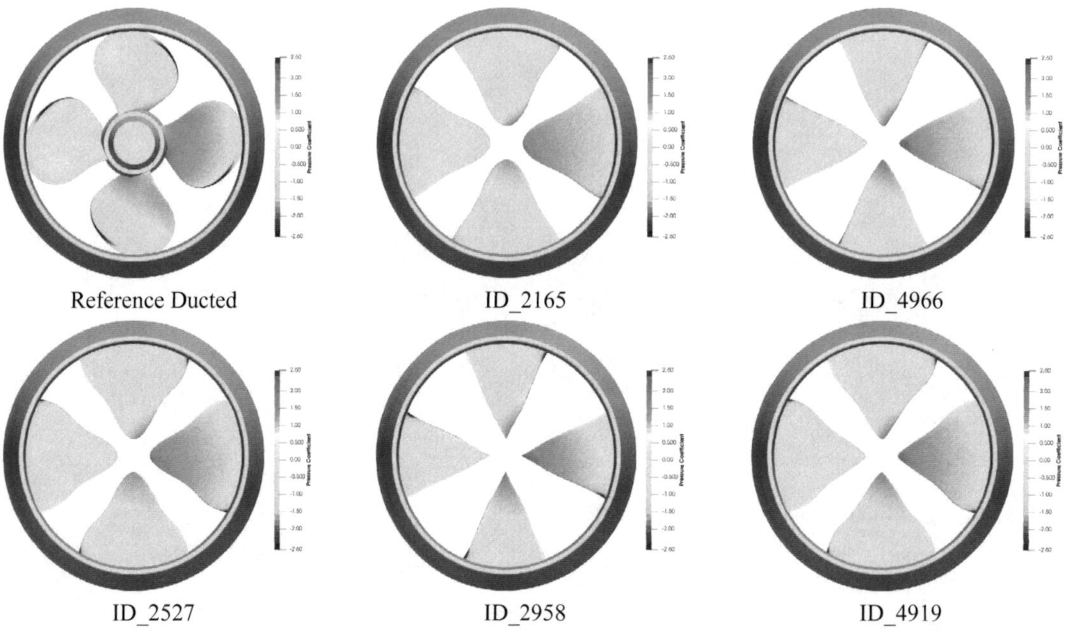

Figure 11: Pressure coefficient distributions on the selected RIM driven propellers. Pressure side.

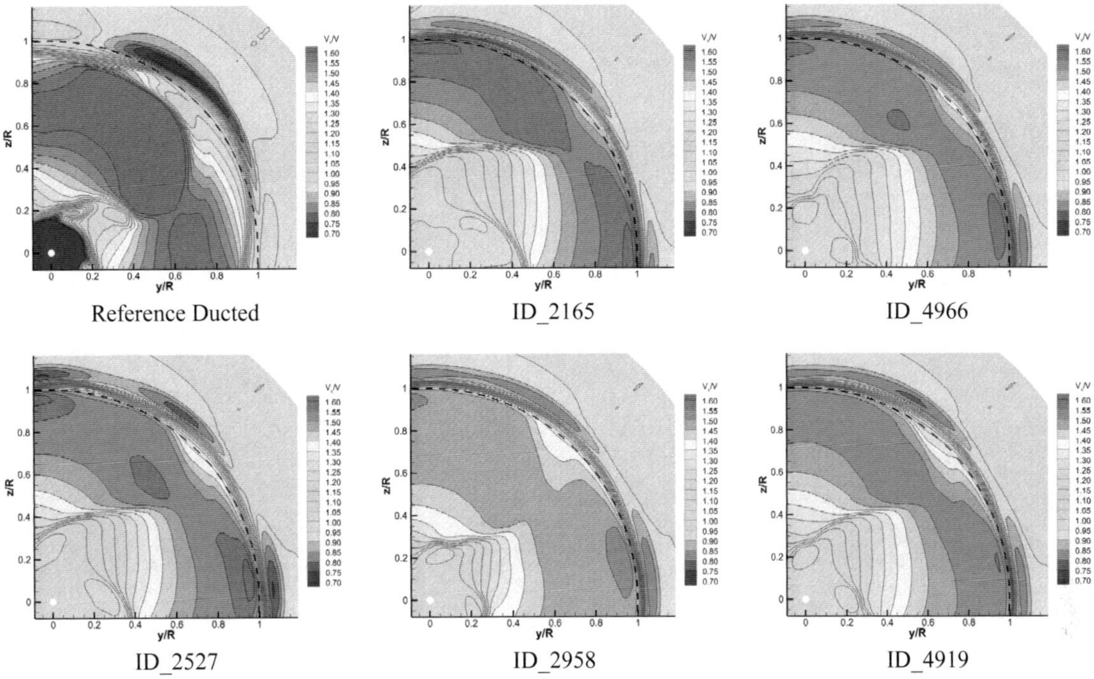

Figure 12: Non-dimensional axial velocity distributions (V_x/V) aft the propeller plane at x/D = 0.5.

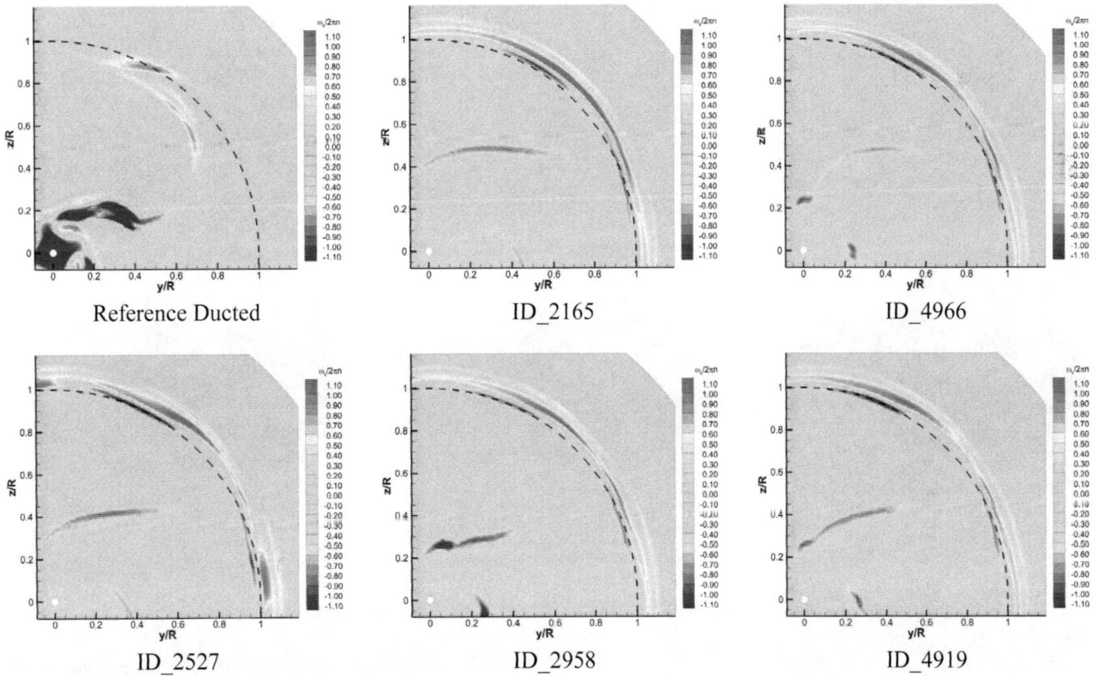

Figure 13: Non-dimensional axial vorticity distributions ($\omega_x/2\pi n$) aft the propeller plane at x/D = 0.5.

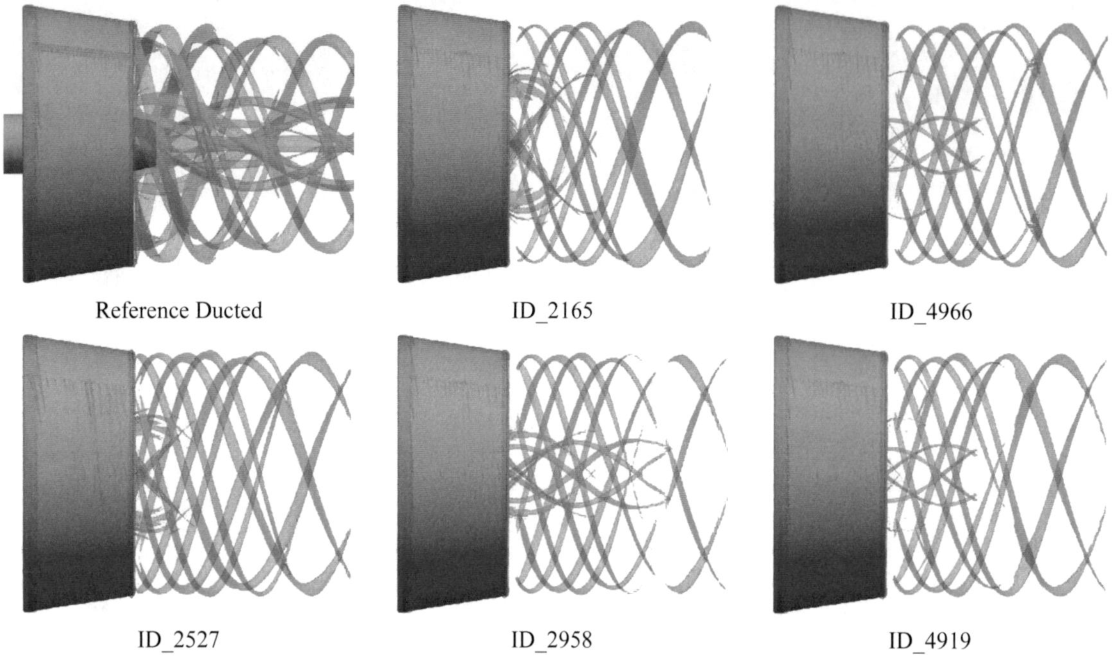

Figure 14: Isosurfaces at constant Q-factor (Q = 15000s^{-2}), longitudinal view.

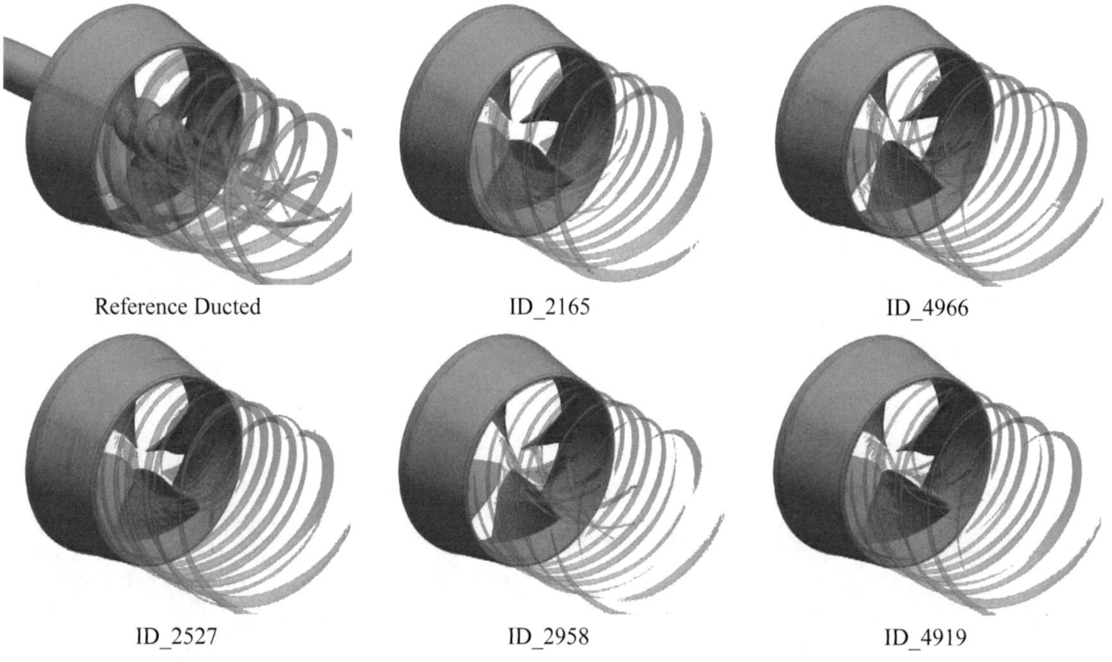

Figure 15: Isosurfaces at constant Q-factor (Q = 15000s^{-2}), view from aft.

The detrimental influence of the "rim" is more relevant the higher is its wetted area and, then, the higher is the chord at the tip that needs to be accommodated in the "rim" itself, contributing to the contrasting nature of the design objectives.

When the attention is moved to the highest efficient RIM driven propellers (ID_4966 and ID_2958), instead, it is possible to appreciate the significant reduction of the expanded blade area which immediately turns into higher values of efficiency (58.5% for ID_2958, then perfectly comparable to that of the reference ducted propeller) and more similar values for cavitation inception on both the suction side (2.80 for ID_2958 versus 3.36 of the ducted configuration) and the pressure side (inception at 3.50 for the RIM, at 4.75 for the conventional ducted propulsor). Phenomena, however, shows some substantial differences: for the reference ducted propeller, suction side cavitation is located at the tip, in the midchord zone, also as a result of the development of the tip leakage vortex developed in the gap region of a blade with finite chord at tip blade, while the risk of leading edge, sheet, pressure side cavitation characterize the midspan of the blade. In the case of RIM propellers, the risk of suction side cavitation is generally higher at the leading edge on the tip zone since the absence of the gap avoids any local vortex development. ID_4966, as well as ID_2958, show a certain inclination to midchord inception which is sustained by certain unloading at the tip. ID_2958, together with ID_4919 and ID_2527, indeed, have the lowest value of pitch at the tip, which determines the higher risk of pressure side cavitation observed for these geometries at the leading edge of the tip zone.

The absence of the gap between the blade and the inner surface of the nozzle significantly changes also the shape of the blade trailing vortex wake, as it can be observed in comparing the non-dimensional velocity and vorticity contours of Figures 12 and 13 or using the isosurfaces at constant Q-factor of Figures 14 and 15. The pitch distributions of the selected RIM driven propulsors is quite "constant" from root to tip while, in the case of the reference ducted propeller, a significant unloading at the tip reduces the local pitch. This difference can be easily recognized by the velocity and vorticity distributions aft the propeller plane (at x/D = 0.5), which show significantly different traces of the vortical wakes. Without the gap, moreover, RIM blades develop only a usual tip vortex, which interaction with the vortexes from the blunt trailing edge of the nozzle determines the grow of a secondary vortex of reduced pitch merging with the primary tip vortex one rotation aft the trailing edge. The primary tip vortex has a comparable strength among the geometries under investigation: only ID_2165 and ID_2527 seem to have slightly more intense vortexes at the tip but their strength is appreciably lower than the vortical structures developed by the conventional ducted propeller.

Further than the tip and the leakage vortex, merged together aft the blades, the conventional ducted propeller is characterized by non-negligible hub vortexes, appreciable from both Figure 14 and by the velocity contour of Figure 12. At hub there are separate vortexes, one for each blade root, traveling far in the wake of the propeller, in addition to the vortex from the hub itself causing the velocity stagnation of Figure 12. RIM configurations, on the contrary, are characterized by very weak vortexes at blade root, which rapidly disappear half a rotation aft the propeller. Only in the case of ID_2958 and, partially, of ID_4966, root vortexes have a slightly higher strength (however decisively lower than that of the conventional ducted propeller), since these two propulsors have the highest pitch at the root. Small chords at root, and consequently lower loads, positively contribute to the strength reduction of these vortexes.

4. CONCLUSIONS

In this paper, we developed a Simulation-Based Design Optimization approach entirely based on OpenFOAM RANS calculations for the design of RIM driven propellers inside accelerating nozzles. The process consists of the automatic handling of thousands of geometries, described using parametric B-Spline curves, from pre-processing and meshing (using cfMesh) to their numerical solution. Computed performance of each candidate, and their compliance with design objectives and constraints drive the convergence towards the Pareto optimal configurations. For the specific case under investigation, we designed a RIM driven propeller able to minimize the risk of any type of cavitation while delivering a given thrust at the design advance coefficient, maximizing at the same time its performance close to bollard pull condition. This problem was formalized as a 9 objective, 18 free parameters optimization, subjected to one constraint, which was solved with the support of a multi-objective genetic algorithm using modeFRONTIER.

The overall best design identified by the SBDO is able to avoid cavitation inception up to a cavitation index of 2.55, which is a substantial improvement with respect to a reference conventional ducted propeller (even if operating at a slightly different functioning condition). Root vortexes, which could represent a side effect of RIM driven configuration, are reduced as well, and contribute to the overall reduction of the vortical structures shed in the wake by the RIM driven configuration.

The cost of these improvements is a slightly lower propulsive efficiency because the reduction of the energy losses due to tip and leakage vortexes partial avoidance is counterbalanced by the substantial increase of the absorbed torque of the rotating "rim". Since some of the effects of the "rim" (the frictional forces in the gap between the rotor and the stator) are neglected due to the simplified model used throughout the optimization, the actual efficiency of the RIM driven propeller could result still slightly lower. Among the feasible geometries, some are able to provide comparable values of efficiency thanks to a sensibly lower expanded blade area. A sensibly higher risk of pressure side cavitation, however, characterizes these configurations. Lower values of pressure can be observed, for these geometries, on the pressure side at the leading edge of the blade in correspondence of the junction with the "rim". This suggests that a simple fillet or higher local leading edge radii (i.e. higher than those currently used in the design process, in particular when chords are particularly wide) can easily dampen the suction peaks and favor higher cavitation inception velocities.

The computational cost of the entire process, finally, is not negligible even if we applied lots of the expedients to speed up the process, from the selection of the most appropriate computational meshes and numerical schemes to the use of a multiple steps workflow devised to save time in case of non-compliance with design objectives and constraints. Once validated, the use of surrogate models and response surfaces in place of expensive RANS calculations could provide a reasonable alternative at least for the very preliminary design phase.

5. REFERENCES

1. SAUNDERS, H. E., 'Hydrodynamics in Ship Design', *Transaction of SNAME*, 1957.
2. LEBEDEV, E. L., PERSHITZ, R. Y., RUSETSKIY, A. A., AVRASHKOV, N. S. and TARASYUK, A. B., 'Ship Steering Units', *Sudostroenie*, 1969.
3. YAKOVLEV, A. Y., SOKOLOV, M. A., and MARINICH, N. V., 'Numerical Design and Experimental Verification of a RIM-Driven Thruster', *Second International Symposium on Marine Propulsors,* 2011.
4. ABU SHARKH, S. M., TURNOCK, S. R. and HUGHES A. W., 'Design and performance of an electric tip-driven thruster', *Proc. Instn. Mech. Engrs Part M, Journal of Engineering for the Maritime Environment, Vol. 217,* 2003.
5. GAGGERO, S., GONZALEZ-ADALID, J. and PEREZ SOBRINO, M., 'Design and analysis of a New Generation of CLT Propellers', *Applied Ocean Research, Vol. 59,* 2016.
6. GAGGERO, S., 'Design of PBCF energy saving devices using optimization strategies: a step towards a complete viscous design approach', *Ocean Engineering, Vol. 159,* 2018.
7. MIZZI, K., DEMIREL, Y. K., BANKS, C. TURAN, O., KAKLIS, P. and ATLAR, M., 'Design optimization of Propeller Boss Cap Fins for enhanced propeller performance', *Applied Ocean Research, Vol. 62,* 2017.
8. MEWIS, F. and GUIARD, T., 'Mewis Duct – New Developments, Solutions and Conclusions', *Second International Symposium on Marine Propulsors,* 2011.
9. BERTETTA, D., BRIZZOLARA, S., GAGGERO, S., VIVIANI, M. and SAVIO, L., 'CPP propeller cavitation and noise optimization at different pitches with panel code and validation by cavitation tunnel measurements', *Ocean Engineering, Vol. 53,* 2012.
10. HUISMAN, J. and FOETH, E. J., 'Automated multi-objective optimization of ship propellers', *Fifth International Symposium on Marine Propulsors,* 2017.
11. GAGGERO, S., TANI, G., VIVIANI, M. and CONTI, F., 'A study on the numerical prediction of propellers cavitating tip vortex', *Ocean Engineering, Vol. 92,* 2014.
12. BALTAZAR, J., FALCAO DE CAMPOS, J. C. A. and BOSSCHERS, J., 'Open-Water Thrust and Torque Predictions of a Ducted Propeller System with a Panel Method', *Second International Symposium on Marine Propulsors,* 2011.
13. LEA, M., THOMPSON, D., VAN BLARCOM, B., EATON, F., FRIESCH, J. and RICHARDS, J., 'Scale model testing of a commercial rim-driven propulsor pod', *J. Ship Production, Vol. 19,* 2003.
14. CAO, Q. M., HONG, F. W., TANG, D.H., HU, F. L. and LU, L. Z., 'Prediction of Loading Distribution and Hydrodynamic Measurements for Propeller Blades in a RIM Driven Thrusters', *Journal of Hydrodynamics, Vol. 24,* 2012.
15. SPARENBERG, J. A., 'On optimum propellers with a duct of finite length', *Journal of Ship Research, Vol. 13,* 1969.
16. MISHKEVICH, V. G., 'Theoretical Modeling of ring propulsor in steady flow', *SNAME Proceedings of Propellers and Shafting Symposium,* 2000.
17. DUBAS, A. J., BRESSLOFF, N. W. and SHARKH, S. M., 'Numerical Modelling of rotor-stator interaction in rim driven thrusters', *Ocean engineering, Vol. 106,* 2015.
18. ESTECO S.R.L., 'ModeFRONTIER 2016 Users' Manual', 2015.
19. THE OPENFOAM FOUNDATION, 'OpenFOAM 4.1.0 Users' Manual', *www.openfoam.org*, 2017.
20. JURETIC, F., 'cfMesh: a novel library for automatic mesh generation', *www.cfmesh.com*, 2017.
21. GAGGERO, S. and VILLA, D., 'Steady Cavitating Propeller Performance by using OpenFOAM, StarCCM+ and a Boundary Element method', *Proc. Instn. Mech. Engrs Part M, Journal of Engineering for the Maritime Environment, Vol. 231,* 2017.

6. AUTHORS BIOGRAPHY

Stefano Gaggero holds the current position of Researcher at the University of Genoa. He is responsible for propulsor flow application and associated development with BEM and RANSE codes.

CFD EVALUATION OF HYDRODYNAMIC PERFORMANCE OF CONTRA-ROTATING PROPELLERS (CRP) USING OPENFOAM

H. Jasak, Wikki Ltd, UK, University of Zagreb, Croatia
L. Balatinec, University of Zagreb, Croatia

SUMMARY

In this paper, a Computational Fluid Dynamics (CFD) study of a contra-rotating propeller (CRP) is conducted, with evaluation of the hydrodynamic performance coefficients of a CRP s et. A series of steady-state (frozen rotor), Harmonic Balance (HB) and transient simulations are carried out using two computational meshes: full propeller geometry and a quarter-geometry with cyclic and partial overlap boundary conditions in a 360° far field m esh. Results of simulations are analysed in terms of hydrodynamic performance coefficients of the CRP s et. The coefficients of the individual propellers are presented and compared with experimental data. Numerical results of the transient and HB simulations show good agreement with the experimental values, while the results of the steady-state MRF simulations show a discrepancy from the experimental value for small advance coefficients, but follow the same trend.

NOMENCLATURE

D_a	Aft propeller diameter (m)
D_f	Fore propeller diameter (m)
J	Advance coefficient
K_{Qa}	Aft propeller torque coefficient
K_{Qf}	Fore propeller torque coefficient
K_{Ta}	Aft propeller thrust coefficient
K_{Tf}	Fore propeller thrust coefficient
$\widetilde{K}_{Qa,N}$	Aft unsteady torque coefficient
$\widetilde{K}_{Qf,N}$	Fore unsteady torque coefficient
$\widetilde{K}_{Q,prop,N}$	Unsteady torque coefficient of a single propeller
$\widetilde{K}_{Q,N}$	Total unsteady torque coefficient
$\widetilde{K}_{Ta,N}$	Aft unsteady thrust coefficient
$\widetilde{K}_{Tf,N}$	Fore unsteady thrust coefficient
$\widetilde{K}_{T,prop,N}$	Unsteady thrust coefficient of a single propeller
$\widetilde{K}_{T,N}$	Total unsteady thrust coefficient
N	N-th harmonic
n	Rotational velocity (s^{-1})
ρ	Water density $(kg\,m^{-1})$
T_a	Aft propeller thrust (N)
$T_{1,N}$	Cosine amplitude component of unsteady thrust (N)
$T_{2,N}$	Sine amplitude component of unsteady thrust (N)
T_f	Fore propeller thrust (N)
Q_a	Aft propeller torque $(N\,m)$
$Q_{1,N}$	Cosine amplitude component of unsteady torque (N)
$Q_{2,N}$	Sine amplitude component of unsteady torque (N)
Q_f	Fore propeller torque $(N\,m)$
$\widetilde{Q}$	Unsteady torque (N)
$\widetilde{T}$	Unsteady thrust (N)
U	Axial velocity (m s^{-1})
η_a	Aft propeller efficiency
η_f	Fore propeller efficiency
θ	Phase angle

1. INTRODUCTION

Conventional blade propellers have been in use since the late 19th century as the predominant means of propulsion for ships, submarines and other maritime vehicles. Due to the fact that single propellers were generally used for propulsion of most ships and submarines, they were the object of a number of studies. More recently, rapid development of Computational Fluid Dynamics (CFD) provides validated and verified predictive capability for propeller performance [1, 2, 3], including their derived characteristics, such as turning circle manoeuvring simulations [4].

With the increased complexity and demand for propulsion, as well as the continued need to design faster, more manoeuvrable and more quiet means of maritime propulsion, contra-rotating propellers (CRP) become an interesting alternative to the conventional single rotor propeller. Carlton [5], described a CRP set consisting of two, usually 4- or 5-bladed, propellers with collinear axes, rotating in opposite directions.

The concept of CRP is mainly used in torpedoes, high-speed outboard units of smaller sizes [6], although after an increased interest in CRP sets and their application in the late 1980s, several examples of CRP being fitted to a bulk carrier in 1988 and a very large crude carrier (VLCC) in 1993 [5] exist. Some older systems tried implementing such a design using two propellers mounted on a single shaft, which resulted in problems with inner shaft lubrication, leading to the stagnation of further research [24].

The idea behind contra-rotating propellers is to position a secondary propeller behind the main propeller, rotating in the opposite direction in order to harvest additional energy otherwise lost in the radial flow. If such a concept is implemented, the secondary (AFT) propeller is fitted to the shaft behind the main (FORE) propeller. An alternative design may be used, such that the aft propeller is fitted to a secondary shaft or pod. This design introduces several improvements over the conventional propeller design [5].

The implementation of CRP reduces the strain exerted on the main propeller by introducing the secondary propeller, which takes over in the production of power needed to generate sufficient thrust to propel the vessel, furthermore a part of the rotational energy is recovered, which would otherwise be transferred to the surrounding water. The two-propeller set reduces the load on individual propellers, also reducing the risk of cavitation along the propeller blades. This results in longer service intervals and an overall increase in lifetime of the propellers. Another benefit discussed by Ghassemi et al. [7] is greater uniformity of the flow filed in the downstream wake of the propeller set, resulting in a lower noise signature.

In recent years, modern variants of the same CRP design have been proposed for a new generation of maritime vehicles. With the development of electric and hybrid propulsion systems, brought on by an increased interest in the reduction of fuel consumption and emissions produced by large transport vessels, the concept of CRP, with the aft propeller mounted on a pod and powered by an electric motor, becomes of great interest.

Studies on the implementation of a podded CRP design [24] show several benefits of using two coaxial propellers rotating in the opposite direction. These include reduction of torque exerted by the propeller on the vehicle, increase in efficiency and decrease in fuel consumption with added benefit of increased manoeuvrability. In case of main engine failure a secondary electric propulsion system can be used. Nevertheless, some disadvantages may be noted. In case of the single shaft design, the aforementioned problem with lubrication of the main shaft may arise with additional maintenance costs. Additionally, the increased cost of the implementation of a secondary power supply for the pod-mounted propeller must be considered.

1.1 PREVIOUS STUDIES

Evaluation of the impact of a CRP set on ship performance has lead to a number of experimental and numerical studies, which will be briefly outlined. A series of experiments using a scale model was conducted by Manen and Oosterveld [11] in 1969 to determine the open water characteristics of a CRP set, while the effects of axial displacement and propeller diameter of CRP were presented by Hecker and McDonald [12] in the 1960. In a later experimental study, Miller [9] to determined the unsteady forces which occur on a CRP set in a uniform flow, by performing water tunnel experiments using a model set. Having encountered difficulties in synchronisation of rotational speed of the two propellers, Miller expands on the previous experiment in 1981, with some success [10]. As mentioned in [7], different authors used various numerical approaches to determine the hydrodynamic performance of CRP sets. Gassemi and Taherinasab [7] used the Boundary Element Method (BEM) to model steady hydrodynamic performance of CRP, Yang et al. [14] and Tsakonas et al. [13] used the lifting surface theory for the same purpose, while Hoshino [15] implemented the same theory while performing experiments for validation. Gu and Kinnas [16] combine the Finite Volume Method (FVM) and the Vortex-Lattice Method (VLM) to evaluate CRP performance. More recently, Wang and Xiong [17] use Computational Fluid Dynamics (CFD) and RANS modelling to investigate the effects of time step size and turbulence models on hydrodynamic performance of CRP sets, while Nouri et al. [18] use CFD to investigate CRP optimisation.

2. METHODOLOGY

Previous research using CFD to investigate the interaction of CRP and calculate hydrodynamic coefficients exists, this paper investigates the capabilities of modern turbomachinery simulation features in `foam-extend`, a well established open-source numerical toolbox for CFD simulations [19].

2.1 MATHEMATICAL MODEL

Three methods were chosen for modelling the relative motion of the propeller set:

- The Multiple Reference Frame (MRF) model, a steady-state frozen rotor approach;
- The Harmonic-Balance (HB) method, a quasi steady-state model;
- A full transient simulation with a sliding mesh.

Each of the methods shall be briefly described below.

2.1 (a) Multiple Reference Frame (MRF) Frozen Rotor

The MRF method [25] is a steady-state method, usually referred to as the frozen rotor approach, where no actual relative motion of the mesh occurs. This model approximates propeller interaction by assigning rotational velocity to specified cell zones in a fixed relative position of the blades. The moving reference frame equations are used to resolve the flow for each of the moving zones, with the equations reducing to their stationary form for stationary zones.

The MRF method does not take into account transient effects and as such cannot be used for rotor-to-rotor interaction. Nevertheless, the MRF method is capable of providing reasonable data of the flow field around the CRP set, especially for operating points in which no large-scale transient effects are present.

2.1 (b) Harmonic Balance Method (HB)

This method is a quasi steady-state method used for simulating temporally periodic flows based on a set of inter-dependent transient temporal snapshots, which act as inter-dependent set of quasi-steady-state problems. Therefore, it is able to describe transient phenomena of the flow [23]. By applying the Fourier series expansion on nonlinear temporally periodic flows described by non-linear transient equations, a set of coupled steady-state equations is used to describe periodic flow over a prescribed set of frequencies through discrete instants in time over a representative harmonic period [26].

The HB method covers both the transient aspects of the flow and the effect of variable relative positions of rotors. Further, its advantage lies in the fact that it reaches the periodic solution without the need to resolve the initial transient.

2.1 (c) Transient Simulations

Transient simulation is conceptually the simplest way of describing the flow in a CRP set, but carries significant disadvantage in terms of computational time. In order to reach temporally periodic solution effects of the initial guess need to be removed from the flow. In practice, this involves simulating a large number of blade passage periods (of the order of 30), with sufficient temporal resolution to capture vortex interaction in the CRP. A practical time-step size measured in rotational angles is approximately 100 time-steps per blade-passage period; for a 4-bladed propeller this would imply 400 time-steps per rotation, which is a significant computational effort.

2.2 INTERFACE MODELLING

In turbomachinery CFD, communication between domains (mesh regions) in relative motion is handled by several types interfaces, depending on the nature of the phenomena observed and the type of simulation model. Two interface models are used: the Mixing Plane interface and the General Grid Interface (GGI), with its variants of partial overlap GGI and cyclic GGI.

2.2 (a) General Grid Interface

The GGI interface [20, 21] enables communication between mesh elements on either sides of non matching surfaces, allowing for non-conforming node location across the patch pair. This is necessary for a pair of regions under relative rotation: it is impractical to perform mesh matching of "clicking" motion at the interface for complex geometries.

The GGI establishes a connection in a conservative and implicit fashion, regardless of node misalignment. It is regularly used in turbomachinery simulations as a way of establishing a connection between unconnected domains. GGI discretisation mimics the polyhedral-type connectivity across the interface at the level of discretisation by applying facet-area-based weights and recalculating cell connectivity for each relative position of a GGI pair.

2.2 (b) Partial Overlap and Cyclic GGI

Two variants of GGI are used in this paper. The first is the overlap GGI interface, which enables communication when full interface coverage is not present in the mesh. Assuming circumferential periodicity, the interface behaves as if the domain was copied a number of times, thus providing a full annulus for the interface communication. The basic discretisation principle re-uses the GGI weight approach, above.

Another variant of GGI is the cyclic GGI interface, used to model non-matching circumferentially periodic patches. Assuming circumferential periodicity, the cyclic GGI is used to establish communication between the patches by applying a rotational transformation as defined by its angle.

2.2 (c) Mixing Plane Interface

In [25, 22] the Mixing Plane model is described as a cost-effective alternative to using a sliding mesh model producing reasonable approximations of the time averaged flow field. In order to circumvent the problem of upstream wakes present in the frozen rotor approach, the Mixing Plane assumes the absence of transient interaction between the upstream and downstream components.

This approach treats each zone as a separate steady-state problem, spatially averaging (mixing) the data from each zone in the radial of circumferential direction (as appropriate) and passing it as a boundary condition, updated in an implicit manner. Such a model removes unsteadiness of flow arising from circumferential variations such as blade passage. Specification of the number of averaging bands and momentum- or area-based averaging is performed depending on the type of averaged variable [8].

2.3 MODEL GEOMETRY

The geometry of the model matches the experiments performed by Miller [9, 10] as closely as possible. The contra-rotating propeller set (type 3686-3687A) consists of two 4 bladed propellers rotating on shafts with collinear axes (Fig. 1, left). The front propeller (type 3686) rotates in the clockwise (CW) direction around the x-axis, while the aft propeller (type 3687A) rotates counterclockwise (CCW) around the same axis. Positive direction of the x-axis corresponds to the axial direction of water flow. Exact specifications of the propeller set used to model the geometry are given in Table 1.

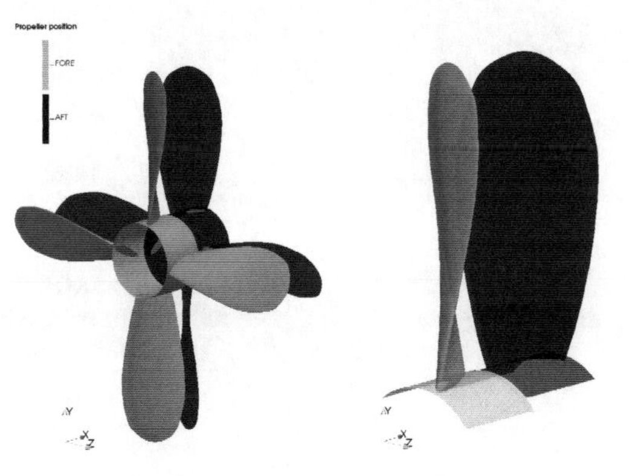

Figure 1: Full (**L**) and quarter (**R**) geometry.

Table 1: Propeller geometry specification.

Propeller type	**3686**	**3687A**
Position in set	FORE	AFT
Blades	4	4
Diameter $[m]$	0.2991	0.3052
Rotation	CW	CCW
Expanded area ratio	0.303	0.324
Section meanline	NACA $a = 0.8$	NACA $a = 0.8$
Thickness Distribution	NACA 66	NACA 66

2.4 COMPUTATIONAL MESH

Based upon previously described geometry, two computational meshes were created. The first mesh contains the complete geometry of both propellers, while the second covers only a single blade passage of each propeller: a quarter of the CRP geometry (Fig. 1, right). The use of such a simplified geometry was possible because of the cyclic nature of the propeller with accompanying phenomena. This enabled the use of the cyclic variation of the GGI in `foam-extend`, with suitable cyclic boundary conditions. The final cell count of the whole mesh was around 5 million cells, whereas the use of the quarter geometry reduced the cell count to around 2 million cells, with both meshes containing only hexahedral cells.

Due to the nature of the mesh used for the numerical simulations, three regions exist which are interconnected by patches (Fig. 2). Two types of interfaces were used for communication between the regions. The communication between the regions surrounding each of the propellers and the outer cylinder bounding the mesh was established with the GGI. For the communication between the two regions surrounding the propellers both the overlap GGI and the Mixing Plane [22] interfaces were tested.

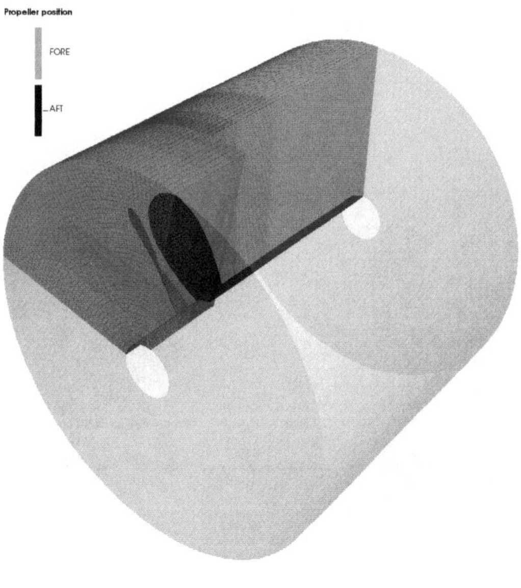

Figure 2: Quarter geometry mesh.

3. HYDRODYNAMIC PERFORMANCE

In this section the numerical data concerning the hydrodynamic characteristic of the CRP set is presented.

3.1 STEADY COEFFICIENTS

Hydrodynamic coefficients describe the characteristics of the fore and aft propellers, as well as the CRP set as a whole:

$$K_{Tf} = \frac{T_f}{n^2 D_f^4 \rho} \qquad K_{Qf} = \frac{Q_f}{n^2 D_f^5 \rho}, \tag{1}$$

$$K_{Ta} = \frac{T_a}{n^2 D_a^4 \rho} \qquad K_{Qa} = \frac{Q_a}{n^2 D_a^5 \rho}, \tag{2}$$

$$\eta_f = \frac{K_{Tf}}{K_{Qf}} \frac{J}{2\pi} \qquad \eta_a = \frac{K_{Ta}}{K_{Qa}} \frac{J}{2\pi}. \tag{3}$$

The advance coefficient J used for both propellers is defined using the diameter of the Fore propeller D_f:

$$J = \frac{U}{n\,D_f}, \tag{4}$$

which permits the use of the following equivalence:

$$K_T = K_{Tf} + K_{Ta} \qquad K_Q = K_{Qf} + K_{Qa}. \tag{5}$$

With the equivalent hydrodynamic coefficients for the whole CRP defined as above, the total efficiency of the set is introduced:

$$\eta = \frac{K_T}{K_Q} \frac{J}{2\pi}. \tag{6}$$

3.2 UNSTEADY COEFFICIENTS

As the data gathered from the simulations needs to be comparable to experimental data, the time-varying results from transient simulation were analysed by the use of the Direct Fourier Transform. As discussed in Miller, [10] the unsteady thrust and torque for the 4×4 CRP set was expected at blade passing frequency (eight times shaft frequency, $N = 8$). The resulting amplitudes were used to determine unsteady thrust and torque coefficients.

Unsteady thrust and torque can be presented as a sum of sine and cosine amplitudes for harmonic N:

$$\widetilde{T}_N = T_{1,N}\cos(N\theta) + T_{2,N}\sin(N\theta), \tag{7}$$

$$\widetilde{Q}_N = Q_{1,N}\cos(N\theta) + Q_{2,N}\sin(N\theta). \tag{8}$$

Unsteady coefficients can thus be defined as:

$$\widetilde{K}_{T,prop,N} = \frac{(T_{1,N}^2 + T_{2,N}^2)^{\frac{1}{2}}}{n^2 D^4 \rho}, \tag{9}$$

$$\widetilde{K}_{Q,prop,N} = \frac{(Q_{1,N}^2 + Q_{2,N}^2)^{\frac{1}{2}}}{n^2 D^5 \rho}. \tag{10}$$

The total unsteady coefficients of the CRP set may thus be presented as:

$$\widetilde{K}_{T,N} = \widetilde{K}_{Tf,N} + \widetilde{K}_{Ta,N}, \tag{11}$$

$$\widetilde{K}_{Q,N} = \widetilde{K}_{Qf,N} + \widetilde{K}_{Qa,N}. \tag{12}$$

4. SIMULATION RESULTS

A series of numerical simulations was performed to simulate different operating points of the CRP set. Both transient and steady-state numerical simulations using the MRF were conducted, with different interfaces connecting the aforementioned mesh regions. The results were analysed and are presented below.

4.1 MESH VALIDATION

An obvious advantage of using numerical mesh of a quarter geometry of the propeller set is a much smaller cell count, which in turn results in lower computational time compared to a full geometry mesh. This fact proves of great significance while running computationally demanding transient simulations. A comparison between the simulation data gathered from a quarter geometry and a full geometry simulation was conducted, showing no loss of accuracy (Fig. 3). This fact confirmed the validity of the quarter geometry mesh and enabled its use in ensuing simulations.

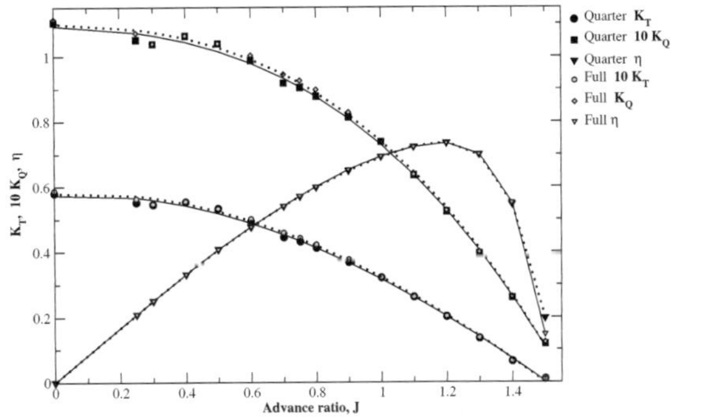

Figure 3: Mesh validation: quarter vs. full mesh.

The data presented in Fig. 3 validates the us of a quarter geometry mesh in further simulations. The aforementioned mesh was adopted for MRF, HB and transient simulations.

4.2 STEADY-STATE RESULTS

A series of steady-state simulations was set-up using two different approaches regarding interface treatment. Both GGI and Mixing Plane interface were used to determine which of these would prove better at connecting the mesh regions without distorting propeller-to-propeller interaction.

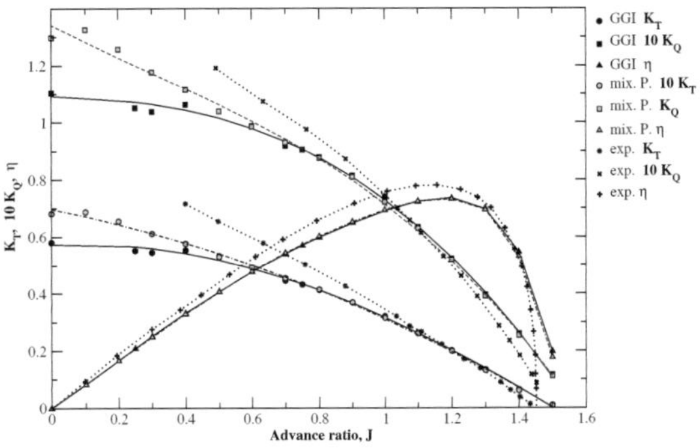

Figure 4: Interface comparison.

Figure 4 shows the comparison between simulations using GGI, Mixing Plane and experimental data gathered by Miller [9, 10] for the CRP set. As expected, the steady-state simulations, regardless of the interface used, deviate from the experimental results, due to significant propeller-to-propeller interaction effects. The deviation increases as the advance coefficient J decreases. This can by explained by the fact that with the decreasing axial velocity, the flow moves from the Best Efficiency Point (BEP) farther into the lower advance coefficient ranges, where the importance of transient phenomena increases. The significantly transient nature of the flow cannot be correctly described by a steady-state frozen rotor approach. Nevertheless, steady state results near the BEP show good agreement with experimental data, which can prove useful as the hydrodynamic performance in this region is usually of primary interest.

Table 2 lists the calculated hydrodynamic performance coefficients for two advance coefficients: $J = 0.5$ and $J = 1.1$. These two points were chosen as to demonstrate the influence transient phenomena have on the accuracy of the steady-state simulations of CRP. The data for both operating conditions was obtained from the of numerical simulations performed on the same quarter geometry mesh, using the overlap GGI and cyclic GGI interfaces, with the segregated SIMPLE-based CFD solver. An analysis of the data presented shows that the accuracy of steady-state results increases, as we move from lower to higher values of the advance coefficient, and the relative error compared to the experimental data diminishes. Thus, moving farther from the area of pronounced transient phenomena if favourable in term of MRF. If such phenomena are to be investigated, a transient simulation should be used.

Table 2: Steady hydrodynamics coefficients for advance coefficients J=0.5 and J=1.1.

	J = 0.5		J = 1.1	
	Value	Error [%]	Value	Error [%]
K_{Tf}	0.2161	37.24	0.1101	9.82
K_{Ta}	0.3168	12.80	0.1515	0.76
K_T	0.5328	22.71	0.2616	4.57
K_{Qf}	0.0378	52.10	0.0259	25.71
K_{Qa}	0.0659	-6.50	0.0375	-14.23
K_Q	0.1389	14.93	0.0634	2.07

By comparing the steady state hydrodynamic performance of the CRP set for two operating points (Table 2), a significant influence of transient on the solution can be seen for $J = 0.5$, whereas the results for $J = 1.15$ show smaller error in comparison. The hydrodynamic coefficients for the whole CRP set show good agreement with the experimental data. Data for separate propellers shows good agreement with the experiment for the fore propellers, with the aft propeller showing some deviation. It should be noted that both the experimental and the numerical data show higher loads on the Aft propeller, which is attributed to the front propeller wake affecting the Aft propeller.

4.3 TRANSIENT AND HB RESULTS

Transient simulations are computationally significantly more demanding on CPU time than steady state simulations, which in turn makes them impractical for large meshes or complex geometry. This emphasises the need to use steady state solvers which are able to capture a fraction of the transient phenomena in their solution. Nevertheless, transient solvers are needed to fully investigate problems with dominantly transient nature, as demonstrated in the previous section. The relatively high error shown in Table 2 for $J = 0.5$ is attributed to transients. As the data acquired by steady state frozen rotor approach diverges from experimental results, a series of transient simulations was made to verify the results. Due to long computational time only a few operating points were analysed.

In all transient simulations, numerical data for the first full rotation of both propellers is discarded to minimise periodic uncertainly before Fourier transform operations are performed. This is considered a minimum acceptable level; due to CPU time limitations, systematic periodic uncertainty study has not been performed.

The Direct Fourier Transform (DFT) was used to analyse time-dependent data gathered from transient simulations. The results showing thrust and torque of the whole CRP set for one complete rotation are shown in Fig. 5. Time-varying values of thrust were transferred to the frequency domain. Beside the base amplitude (mean value of thrust or torque), the expected appearance of an unsteady component of thrust can be noted at eight times shaft frequency. This was also observed in the experiments performed by Miller [9, 10]. The values of unsteady thrust and torque from the DFT analysis of the transient simulation for $J = 0.5$ have been used to calculate the unsteady hydrodynamic coefficients at $N = 8$, given in Table 3 without comparison to experimental data as such data for $J = 0.5$ was not available.

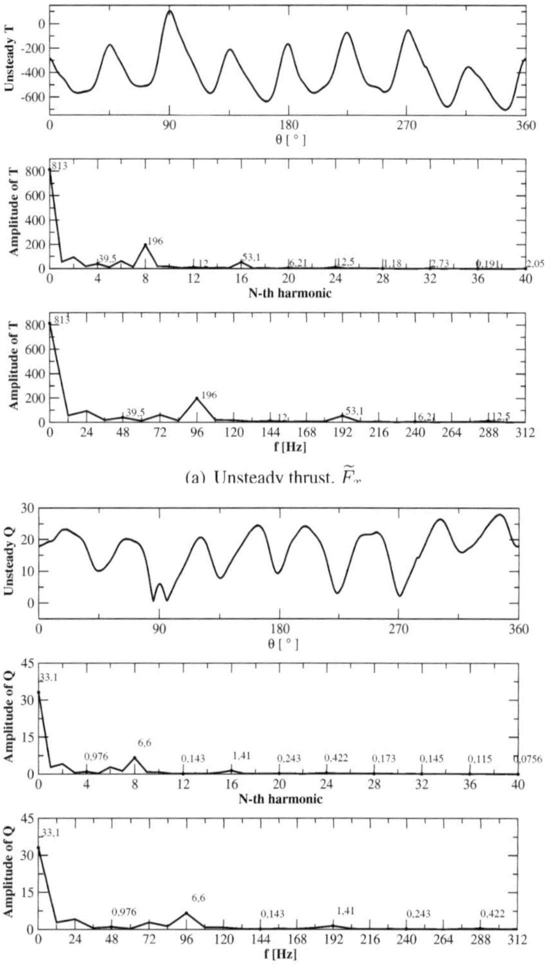

(a) Unsteady thrust, $\widetilde{F_x}$

(b) Unsteady torque, $\widetilde{M_x}$

Figure 5: Transient results for CRP set at $J = 0.5$

The same table shows the results from using the Harmonic Balance method (HB). In comparison to steady-state data, both the transient simulation and the HB method show an increase in accuracy of the results, with the HB method being less accurate than the transient simulation, but having a significantly shorter computation time. The choice of method used should be done based on the accuracy desired.

Table 3: Simulation data for J=0.5

	Experiment	Transient		HB	
	Value	Value	Error [%]	Value	Error [%]
K_T	0.6538	0.6891	-5.12	0.5970	13.3
K_Q	0.1194	0.1170	2.10	0.1139	4.6
η	0.4357	0.4689	-7.07	0.4170	4.4
$\widetilde{K}_T$	-	0.1655	-	-	-
$\widetilde{K}_Q$	-	0.0284	-	-	-

5. CONCLUSIONS

The work presented in this paper demonstrated the validity of numerical simulations of Contra-Rotating Propellers using OpenFOAM. Several different approaches regarding the computational mesh and geometry, interface handling and type of simulation were investigated. The results were presented in terms of hydrodynamic coefficients.

The quarter geometry mesh was chosen for its size and the resulting increase in computation speed. The mesh was validated against its whole geometry counterpart showing good agreement between simulations. The choice of mesh handling was investigated, opting for using a combination of variants of the General Grid Interface (GGI): the cyclic GGI and overlap GGI.

Results of steady-state simulations for different advance coefficients were compared to transient and Harmonic Balance method simulations. The use of steady-state solvers is justified for higher values of the advance coefficient J, resulting in good agreement with experimental data. This can be justified by the fact that higher values of J, meaning higher values of the axial velocity, occur for higher efficiency operating points where the transient phenomena are less pronounced. On the other hand, the lower values of J coincide with a more pronounced influence of transients. The use of transient solvers is necessary to achieve satisfying results for this area, with an added capability of unsteady forces/torque analysis not possible when using steady-state solvers. An alternative to using transient solvers is the Harmonic Balance method, giving results close to those of the transient simulations in a fraction of the computational time. Both methods produce results with satisfactory conformance to experimental data.

6. REFERENCES

[1] Jasak, H., Vukčević, V., Gatin, I. & Lalović, I., 'CFD validation and grid sensitivity studies of full scale ship self propulsion', International Journal of Naval Architecture and Ocean Engineering, 2018

[2] Krasilnikov. V. I., 'Self-Propulsion RANS Computations with a Single-Screw Container Ship' Third International Symposium on Marine Propulsors smp13, Launceston, Tasmania, Australia, 2013

[3] Šeb, B., 'Numerical Characteristics of A Ship Propeller' Master's Thesis, Zagreb 2017

[4] Gatin, I., Vukčević, V., Jasak, H.& Lalović, I., 'Manoeuvring simulations using the Overset grid technology in foam-extend', 32nd Symposium on Naval Hydrodynamics, Hamburg 2018

[5] Carlton, J. S., 'Marine Propellers and Propulsion, 3rd ed.', Elsevier, 2012

[6] Brizzolara, S., Tincani E.P.A. & Grassi, D., 'Design of contra-rotating propellers for high-speed stern thrusters', Ships and Offshore Structures, 2007

[7] Ghassemi, H. & Taherinasab, M., 'Numerical calculations of the hydrodynamic performance of the contra-rotating propeller (CRP) for high speed vehicle', Polish Maritime Research, 2013

[8] De Dominicis I., Cvijetić G., Willetts M., Jasak H. 'Capabilities for foam-extend: Development and Validation', In Nóbrega J., Jasak H.: OpenFOAM: Selected Papers of the 11th Workshop, Springer, doi: https://doi.org/10.1007/978-3-319-60846-4, 2019

[9] Miller, M. L., 'Experimental Determination of Unsteady Forces on Contrarotating Propellers in Uniform Flow', David W. Taylor Naval Ship R & D Center, Bethesda Md., 1976

[10] Miller, M. L., 'Experimental Determination of Unsteady Forces on Contrarotating Propellers for Application to Propellers', David W. Taylor Naval Ship R & D Center, Bethesda Md., 1981

[11] Van Manen, J. D. & Oostervald, M. W. C., 'Model Tests On Contra-Rotating Propellers', Netherlands, 1969

[12] Hecker, R. & McDonald, N. A., 'The Effect of Axial Spacing and Diameter on The Powering Performance of Counterrotating Propellers', David W. Taylor Naval Ship R & D Center, Bethesda Md., 1960

[13] Tsakonas, S., Jacobs, W. R. & Liao, P., 'Prediction of steady and unsteady loads and hydrodynamics forces on contra-rotating propellers', Journal of Ship Research, 1983

[14] Yang C. J., Tamashima M., Wang G. Q. & Yamazaki R., 'Prediction of the steady performance of contra-rotating propellers by lifting surface theory', Transactions of the West-Japan Society of Naval Architects No. 82, 1991

[15] Hoshino, T., 'Experimental and Theoretical analysis of propeller shaft forces of contra-rotating propellers and correction with full scale data', Propeller/Shafting94 Symposium, Society of Naval Architects and Marine Engineers, Virginia Beach, 1994

[16] Gu, H., Kinnas, S.A., 'Modelling of contra-rotating and ducted propellers via coupling of a vortex-lattice with a finite volume method', Propellers/Shafting 2003 Symposium, Society of Naval Architects and Marine Engineers, Virginia Beach, 2003

[17] Wang, Z. & Xiong, Y., 'Effect of Time Step Size and Turbulence Model on the Open Water Hydrodynamic Performance Prediction of Contra-Rotating Propellers', China Ocean Engineering, 2013

[18] Nouri et al., 'Optimization of a marine contra-rotating propellers set', Ocean Engineering, 2018

[19] Jasak, H., 'OpenFOAM: Open source CFD in research and industry', International Journal of Naval Architecture and Ocean Engineering, 2008

[20] Beaudoin, M. & Jasak, H., 'Development of a Generalized Grid Interface for Turbomachinery simulations with OpenFOAM', Open Source CFD International Conference, Berlin, 2008

[21] Jasak, H., 'General Grid Interface Theoretical Basis and Implementation', NUMAP-FOAM Summer School, Zagreb, 2009

[22] Beaudoin, M., Nilsson, H., Page, M., Magnan, R., Jasak, H., 'Evaluation of an improved mixing plane interface for OpenFOAM', The International Association for Hydro-Environment Engineering and Research, 2014

[23] Cijetić, G., 'Harmonic Balance Method for Turbomachinery Applications', 2nd FAMENA PhD Workshop, 2016

[24] Van Kluijven, P. C. et al., 'Contra-rotating propellers', Rotterdam Mainport University of Applied Sciences RMU

[25] Cijetić, G., 'Steady State Methods For Turbomachinery', Faculty of Mechanical Engineering and Naval Architecture, Zagreb, 2014

[26] Cijetić, G., 'Analysis and Implementation of the Harmonic Balance Method in Computational Fluid Dynamics', Master's Thesis, Faculty of Mechanical Engineering and Naval Architecture, Zagreb, 2014

7. AUTHORS BIOGRAPHY

Hrvoje Jasak is the co-author and leading developer of OpenFOAM, an open source object-oriented C++ library for numerical simulations in continuum mechanics. In the period of 1993 to present, he has been driving the FOAM/OpenFOAM development effort, from PhD at Imperial College, researching numerics at CD-adapco, as the Technical Director at Nabla Ltd. and at Wikki Ltd. He initiated the OpenFOAM Workshop conference series, as well as NUMAP-FOAM summer school. He became an associate professor in 2007 at the University of Zagreb, Croatia and was promoted to the rank of full professor in 2016.

Luka Balatinec is a second year Master's student at the University of Zagreb, where he is currently working on his MS degree. The topic of his Master's thesis is An Overview of Rotor-Stator Interfaces for CFD Simulations in Turbomachinery, focusing on the hydrodynamic performance of contra-rotating propellers.

PROPELLERS GEOMETRY MODELING THROUGH B-SPLINES SURFACES

Rodrigo Perez Fernandez, Universidad Politécnica de Madrid/Spain, rodrigo.perez.fernandez@upm.es
Francisco Pérez-Arribas, Universidad Politécnica de Madrid/Spain, francisco.perez.arribas@upm.es

SUMMARY

This work presents a new design methodology for modelling the blades of ship propellers using B-spline surfaces that are a standard tool in CAD and Naval Architecture software products.
Propeller blades of a ship are good examples of free form surfaces, designed specifically considering several parameters that control their performance. Traditional tools for surface design in CAD such as control point manipulation, are not appropriated for blade design, and the designers prefer to work with a collection of propeller parameters that ultimately represent its surface and that possess a clear hydrodynamic meaning.
This method uses common design parameters for the geometry of propellers and produces a final B-spline surface for the geometry of the blades that can be used for the visualization, calculations, and construction of the propeller.
The method starts with the definition of a 3D grid of points that form the propeller blades based on the 2D definition of a series of cross-sectional profiles at several radial locations. These 3D points consider the inclination and twist of the blades are given by rake and pitch angles, quite common in the design procedures, and also different propellers parameters such as skew and blade thickness distribution.
Propeller blades are very thin objects with great changes of curvature, and if standard B-spline techniques are used, they cannot be modelled well under a tolerance unless a large number of control points is used, producing very complex surfaces.
The method stresses the fitting of the blade's leading edge which has great effect on the propeller behavior and geometrically has a small curvature radius in comparison with the rest of the blades. The leading edge is difficult to reproduce with standard techniques.

NOMENCLATURE

CAD	Computer Aided Design
CFD	Computer Fluid Dynamics
ILS	Iterative Least Squares
LS	Least Squares

1. INTRODUCTION

Blades of a ship propeller are examples of functional free form surfaces that are designed by naval architects for a practical purpose, so the ship can advance at a desired speed with a given power.
Geometrically, a blade is typically defined by a series of cross-sectional profiles stacked at several radial locations. The blade shape of a propeller is defined by a series of cylindrical cross-sections at specified radius ratios. The inclination of the blades in respect to the XZ plane is given by the rake distribution and there are other specific parameters, such as skew and pitch distributions. In this figure, a left-hand ship propeller is presented, where the propeller reference plane is the XZ plane and the flow direction is along the Y axis, which is also the rotation axis. The hub is depicted as a cylinder and the base circles are presented. The blade sections of that figure are contained in concentric cylinders which are defined to represent the path of fluid through the propeller.
The blade design and its geometrical definition are intrinsically related to the selection of 2D profiles with good hydrodynamic performance. Therefore, the geometrical characteristics of the 2D sections must be accurately modelled in a 3D definition of the propeller blades.

So, the approach presented for defining the geometry of the propeller blades is based on a discrete approximation of a collection of space curves or cross sections, containing a 3D representation of the 2D profile points. The manner in which these points are obtained and the tolerances associated with the data, influence the presented method for the representation of the blades. This blade definition corresponds to the classical construction of a propeller based on a generator line and profiles attached to it. A classical ship's propeller lines drawing is shown in Fig. 1, where the expanded or unrolled 2D profiles are presented at different stations, and two projections are used to represent the complete propeller geometry. Some propeller manufacturers still use this kind of representations to construct ship propellers at full scale, using well established manual techniques that are based on a set of patterns that experience has demonstrated that work well in practice.

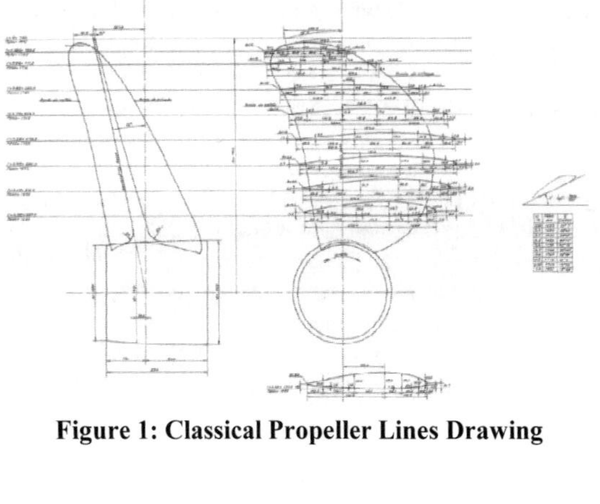

Figure 1: Classical Propeller Lines Drawing

Modelling the geometry of a propeller is different from modelling other industrial objects; for example, an accurate representation of the shape of the sections is required, and this requires working with a large amount of information because the representation of these curves is based on a discrete set of data points, usually about 200 points per curve. If standard techniques are used, very complex surfaces with a large number of control points are obtained and will likely present poor smoothness in the resulting surface obtained from the interpolation of the data points (Perez-Arribas et al. 2018) [1].

The results defined by the blade design process have to be compatible with existing standard technology in CAD, and are difficult to solve by direct or manual manipulation of the control points in the surfaces. B-spline representation of blade surfaces is widely used, and enable subsequent steps in the product development such as manufacturing, finite element analysis, or *Computer Fluid Dynamics* (CFD) calculations. Techniques for propeller design have a long history with the first scientific works published in the 1960's. If the reader would like to know more about hydrodynamic design of propellers two good starting points are the readings in references (Carlton, 2007) [2] and (Adkins, 1994) [3].

In more recent years, the use of CAD methodologies and software products, opened new possibilities for blade design. As mentioned, blades are typically defined by a series of cross-sectional profiles stacked at several radial locations. These sections can be placed on planes in the case of wings of foils, on concentric cylinders in the case of propellers, or in concentric cones in the case of turbo machinery (gas, steam, and Pelton turbines mainly). Thus there is a wide scope for blade representation in the literature, with common geometry basics, and there are different techniques for convenient shape modifications of the blade surfaces. Reference (Ye and Jackson, 1996) [4] presents a technique for designing functional surfaces including normal vectors as physical constraints of the surface. The method focused on propeller design based on hydrodynamic analysis but did not use B-spline representation. In reference (Hoschek and Müller, 1999) [5], turbine blades are first reproduced via lofted B-spline surfaces from a set of given planar profile curves, also produced in B-spline representation. This reference does not consider how the profile curves are obtained, which is a key factor in propeller design. Reference (Ye and Jackson, 1996) [4] presents an approach for turbine blade that allows the designer to control certain geometric parameters of the blades, and include blade material. However it does not consider the use of sections and needs an initial surface definition. Reference (Hoschek and Müller, 1999) [5] uses B-spline curves to represent turbine blades, based on 2D blade sections. This reference uses a large number of data points to model the 2D sections, and uses the same points as the B-spline control points. It will be seen later that a reduced number of control points is desired, since a large number of control points will likely present poor smoothness in the resulting surface. This reference also used the lofting technique, so the blade can be produced with a single surface.

An interesting reference is (Qian and Dutta, 2001) [6] that is focused on the modelling of propellers. In these reference, different aspects of propeller modelling with B-spline surfaces are presented describing how PropCad Software developed by Hydrocomp deals with propeller 3D representation. There is little detail on how the sections are modelled, because the software uses Hydrocomp propriety spline routines, but the paper presents best practices that can be used when lofting the propeller sections.

The first step of the presented method is to accurately model the 3D data points of propeller sections, with special attention to their leading edge. An iterative least squares fitting process ((Koini et al., 2009) [7]) with an original selection of the parameterisation based on the minimum of the maximum Euclidean distances, (or Haussdorf metric) is described in Section 3 and also considers accurate modelling of the leading edge.

The proposed method allows a reduction in the number of control points of the profiles without reducing accuracy. Data reduction in an object representation speeds up most of the downstream processes, improves the fairing of the resulting surface and decreases storage requirements in following design stages. Once the profiles have been accurately modelled with B-spline curves, the next step is to construct a surface that contains them. The technique of lofting or skinning over B-spline curves was selected from the various numerical methods used for industrial surface definitions. This consists of fitting a surface through an ordered set of cross-sectional curves (Kaplan and Macpherson, 2014) [8], (Piegl and Tiller, 1997) [9], which in this case are the propeller profiles.

The paper outline is described as follows; first, the presented method reads the geometric information of the blades, including the position, skew, rake and twist distribution and the discrete data definition of the sections, which the designer has determined after the initial hydrodynamic design stage. All the available inputs are geometric by nature and are familiar for a naval architect. A discrete 3D definition of the profiles is then defined in Section 2. A transformation is developed which allows construction of, and interaction with, 2D blade sections that are mapped onto cylindrical surfaces (fluid stream surfaces). Therefore changes of these planar curves and parameters, result in immediate changes to the 3D propeller model.

It follows the definition of a knot vector that adapts well to the discrete 3D section shape, producing a non-uniform knot vector (Section 4). Afterwards an iterative curve fitting scheme is applied to the discrete data, resulting in a B-spline curve with its control points (Section 6). This fitting considers a fairing parameter in order to obtain realistic curves, since the problem is strongly non-linear and can produce unfaired solutions. The method includes also a way to include more control points near the leading edge as explained in section 5. This fitting is tested considering different 2D shapes that are used in marine applications in Section 8.

Finally, a lofting surface of the previous curves is applied, following the span-wise direction along which the propeller sections are positioned. One application example is used to evaluate the practical usage of the presented method. A graphical outline of the full method is presented in Fig. 2.

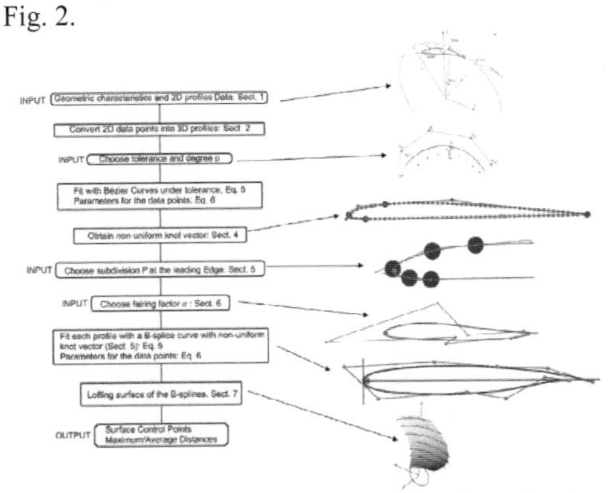

Figure 2: Flowchart of the proposed method

2. DISCRETE SURFACE DEFINITION: BLADE SECTIONS

For many blade sections its 2D geometry is mathematically derived, and the data is put into an ordinate file. These files of data points can be obtained from software or from data bases and they can contain a higher density of points in the leading edge, in order to accurately reproduce this important part of the profile. The number of data points is usually over 200 points per section.
The first step is to create a 3D definition of the blade sections based on the 2D points, so they can be fitted with B-spline curves. The stacked cross sections define the blade.

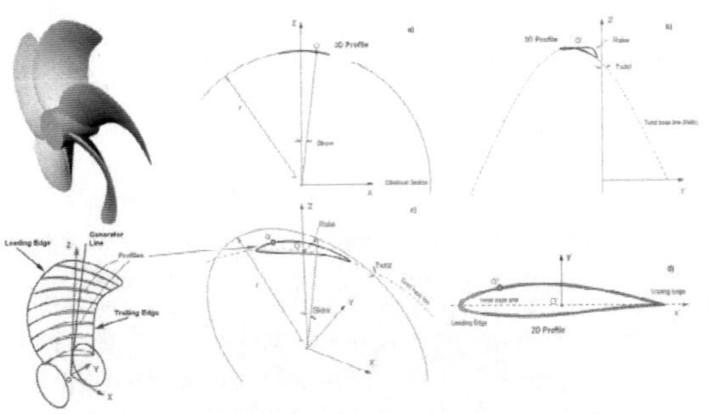

Figure 3: Geometry of the blade sections

Considering the reference system as in Fig. 3, the base line of each aerofoil section is part of a helix line, as depicted in Fig. 3b) and c), which winds on the surface of a cylinder of radius *r*. The twist angle varies along the span of the blade, and the twist is defined as the angle between the section base line and the circle of radius *r* inside the propeller plane that corresponds to the section, as shown in Fig. 3c). This circle is also the intersection of a cylinder of radius *r* that contains the profile with the XZ propeller plane. Notice that a cylinder is used in the case of propellers because it approximates well to the stream surface that contains the fluid flow. In the case of wings or foils, stream surfaces would be planes, and cones in the case of turbo-machinery.
The rake is defined for each section as the distance from the profile origin O' to the propeller plane, as depicted in Fig. 3b) and c). Usually, the rake of each section can be defined as a function of a constant angle for the whole blade, so the rake of each section is a linear function of this angle.
The skew angle is the angle between O' and the Z axis, and it controls the shape of the blade, as shown in Fig. 3a) and c). This angle is obtained based on the generator line that crosses through the reference points O' of the sections. It can be a straight line or a soft curve according the design characteristics.
The origin O' of the profiles can be placed well within the chord midpoint (Fig. 3d), at the point of maximum thickness or at the leading edge. Therefore, its position affects the value of the skew distribution.
All of the above-mentioned geometric parameters, together with the number of blades, diameter and offset tables to define the profiles at specified radii, are required to define the geometry of the propeller, as presented in the drawing of Fig. 1. This data is used to generate the 3D coordinates of points contained on the surfaces of the blades placed at different radius ratios (stations). In this way, the coordinate points of the surface of a single blade are obtained. The rest of the blades are formed by a homogeneous coordinate transformation of a single blade about the axis of rotation. The hub is modelled in its simplest form as the frustum of a right circular cone.
In this paper, a variable presented in **bold** letters indicates this element is formed from several components, such as points, vectors or B-spline curves may have *X*, *Y* and *Z* components. The rest of the variables will be noted in *italics* inside the text. Thus, the 3D coordinates of any point $\mathbf{Q_i}(x,y,z)$ can be obtained based on their associated 2D coordinates $\mathbf{Q_i}'(x',y')$ following these steps (see Fig. 4 to be aligned with the notation as follow):

1) The twist angle of each section Tw_i is applied, and $\mathbf{Q'_1}$ is obtained:

$$R = \sqrt{x'^2 + y'^2}$$
$$\theta = \mathrm{atan}(y'/x') \tag{1}$$
$$\mathbf{Q'_1} = \left(R\cdot \mathrm{Cos}(\theta + Tw_i), R\cdot \mathrm{Sin}(\theta + Tw_i)\right)$$

2) The point $\mathbf{Q'_1}$ is rolled over the surface of the cylinder or radius r_i of the ith section to obtain $\mathbf{Q_2}$. This bending sets the arc length over the cylinder to be equal to the abscissa of $\mathbf{Q'_1}$:

$$\alpha_i = \frac{x'(\mathbf{Q'_1})}{r_i} = \frac{R\cdot \mathrm{Cos}(\theta + Tw_i)}{r_i} \tag{2}$$
$$\mathbf{Q_2} = (r_i\cdot \mathrm{Sin}(\alpha_i), R\cdot \mathrm{Sin}(\theta + Tw_i), r_i\cdot \mathrm{Cos}(\alpha_i))$$

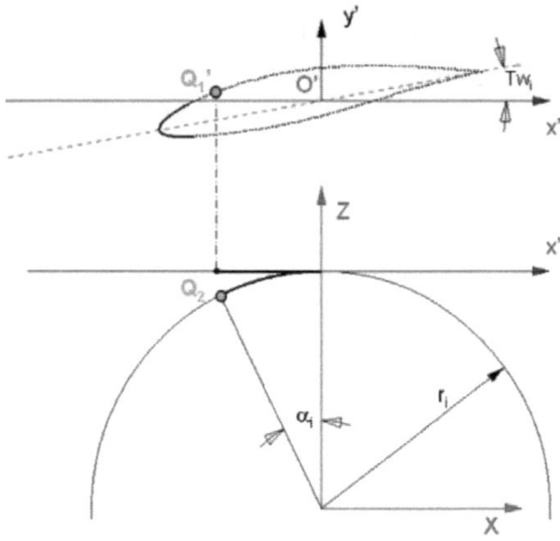

Figure 4: Twisting and bending

3) The final step applies the skew (Sk_i) and rake (Rk_i) at a radius r_i to the bent points by adding these magnitudes to the angle α_i and to the Yth coordinate, respectively. These additions maintain the points over the surface of the cylinder of radius r_i, and vary along the radius:

4)

$$\mathbf{Q}_j = (r_i \cdot \mathrm{Sin}(\alpha_i + Sk_i), R \cdot \mathrm{Sin}(\theta + Tw_i) + Rk_i, r_i \cdot \mathrm{Cos}(\alpha_i + Sk_i)) \tag{3}$$

By repeating this procedure for all points (j=0, $np(i)$) of each profile (i=0, M), a discrete approximation of the blades by a set of data points is obtained. The next step is the modelling of these discrete curves as B-splines so the final lofting can be obtained.

3. FITTING DATA POINTS WITH B-SPLINES USING THE LEAST SQUARES TECHNIQUE

For the sake of simplicity in the notation, it will be considered in the following sections that all the propeller sections i will be defined with the same number of data points, so $np(i) = np$. Graphics during the explanation of the method show 2D curves looking for clearer and more uniform figures, but the method works the same considering 3D curves and points.
Once the $np + 1$ discrete 3D data points $\mathbf{Q_j}$ have been obtained, the B-spline fitting can be conducted. A large number of data points do not suggest the use of an interpolating B-spline. Instead, an approximating curve $\mathbf{c(t)}$ is required. This curve does not pass exactly through the data points but close enough to the points to capture their inherent shape. This is the well-known *Least Squares* (LS) approximation, discussed in (Koini et al., 2009) [7].
In a general fitting problem, np+1 data points $\mathbf{Q_0}, \ldots \mathbf{Q_{np}}$ are approximated by a B-spline of the pth degree with N+1 control points $\mathbf{P_0}, \ldots \mathbf{P_N}$, $N < np$ that are unknown and are obtained as the final results of the calculations, see Fig. 5.
The general LS problem is described by an over-determined set of np+1 equations with N+1 unknown variables:

$$\sum_{i=0}^{N} B_i^p(t_j) \cdot P_i = Q_j, \quad j = 0, \ldots, np \tag{4}$$

Where B_i^p corresponds to the ith basis function of a pth degree B-spline that is calculated using de Boor's algorithm, and considering a uniform knot vector. In (4), t_j (j=0, n) represents the parameters associated with the data points. The graphical meaning of these t_j that is used in this paper, is depicted in Fig. 5.

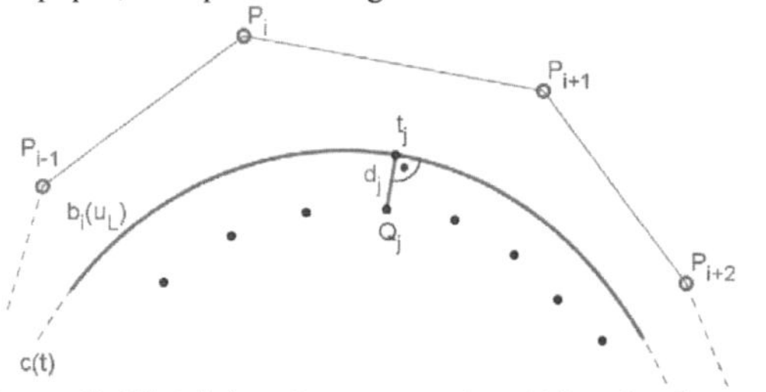

Figure 5: Obtaining the parameters tj for the data points fitting

Matrix expressions are convenient to solve the problem:

$$\sum_{i=0}^{N} B_i^p(t_j) \cdot P_i = Q_j, \; j = 0, \ldots, np; [M] \cdot [P] = [Q] \Rightarrow [M]^T \cdot [M] \cdot [P] = [M]^T \cdot [Q] \tag{5}$$

This system of (np+1) by (N+1) equations is solved by multiplying both sides of Eq. (5) by $[M]^T$, which creates a determined (N+1) by (N+1) linear system.
This type of system can be poorly conditioned, especially if a large number of control points are used. A conventional technique should not be used to solve this ill-conditioned system. Instead, a single-value decomposition of $[M]^T \cdot [M]$ and a later back-substitution process are performed. The solutions of this system are the control points of the best B-spline fitting.
Up to now, a standard LS fitting has been described. Approaching this problem with a standard parameterization such as centripetal or chord-length parameterization is correct but does not consider the effect of the distance of the data points to the B-spline. In this method, parameterization based on a minimum distance metric is adopted (Haussdorf metric). The process is iterative and is described by the following three steps:

1. The method begins with a centripetal parameterization of the $\mathbf{Q_i}$ points, and system (5) is solved. This produces a starting curve of the iterative process only for the first loop, so no initial guess is needed.
2. For each $\mathbf{Q_j}$, the minimum distance to the B-spline is calculated, see Fig. 5, which leads to a solution to Eq. (6).

$$(Q_j - c_i(t_j)) \cdot (c'_i(t_j)) = 0 \tag{6}$$

Once the solution has been found, this t_i value is the parameter associated with the point $\mathbf{Q_i}$ when solving system (5). This is graphically described in Fig. 5.

3. After obtaining the t_j (j=1, np) values, the distance d_j = ($\mathbf{Q_j}$ – $\mathbf{c}$(t$_j$)) is computed, which is the Euclidean

distance between $\mathbf{Q}_i$ and the B-spline. This distance is used to verify the shape requirements. If the maximum distance d_j (j=1, n) is above a given tolerance, steps 2 and 3 are repeated until an acceptable maximum distance is obtained. The quality of the obtained curve is measured using the tolerance constraint, and the shape of the B-spline is amended using parameterisation (6).

Hitherto, the *Iterative Least Squares* (ILS) of a set of points has been described. This technique is used first to determine the best knot vector for the B-spline that fits the profile data points, and once this knot vector has been obtained, a new stage of the ILS is used to obtain the final B-spline. These two steps are described in the following sections.

4. OBTAINING THE DEGREE AND THE KNOT VECTOR

A generic B-spline of the pth degree and with NP control points, is composed of $NP - p$ Bézier curves of the same p^{th} degree, with p+1 control points each. The joining points of the Bézier pieces are called knots and they define the knot vector. Standard fitting techniques use uniform knot vectors. There are other methods to obtain non-uniform vectors, but most of the methods that optimize the knot vector are complicated and very time consuming due to the optimization process that they use. The text book (Woodward, 1987) [10] describes the use of non-uniform knot vector using optimization techniques.

The present method produces fitting of propeller curves under a given tolerance by playing with the knot vector without using optimization techniques, which is selected in a non-uniform way following these steps:

1. Introduce data point one be one: starting in the trailing edge $\mathbf{Q}_0$, data points are introduced sequentially maintaining the number of control points of the Bézier curves equal to p+1, first solving the system (5) and then obtaining the maximum distances accordingly (6).
2. Check the tolerance: when a new data point $\mathbf{Q}_k$ is introduced, the maximum distance is computed and checked with the tolerance value. If the maximum distance is below the tolerance, the following point $\mathbf{Q}_k$+1 is introduced, and equations (5) and (6) are again computed considering the new extra data point. If the maximum distance is over the tolerance, which means that a new Bézier piece should start, as depicted in Fig. 6. In this figure, the large black dots mark the start and end of the mentioned Bézier pieces. The figure shows the different p = 3 Bézier pieces for a 2D profile with 100 data points and one meter chord, with a flat face that is quite common in marine propellers. Every piece will have p + 1 = 4 control points. In this case the profile can be modelled in 4 pieces with a tolerance tol = 0.001 m. Notice how for the last piece, the 4 control points are mostly aligned, since this profile has a flat aft part. It seems clear how the leading edge will need more pieces than the rest of the profile parts.

Figure 6: Example of knots distribution, p = 3

3. If tolerance is over passed: the following Bézier piece will have the first control point equal to the last one of the previous piece, and the data points are now introduced starting from $\mathbf{Q}_k$+1 and finishing when the maximum distance of the new subset of data points computed with (6) overpasses the tolerance. Fig. 7 shows a detail of the leading edge of the previous profile. This part of the curve will need 2 pieces to fit the points with the tolerance, the second piece fits from point 42 until 50, the third piece from 50 until 56. In this figure, the radius of the circles has been increased to 5 mm, so the numbers inside are clearer. Notice how the Bézier curves present a *C0* continuity between them, as is clear at 50 and 55.

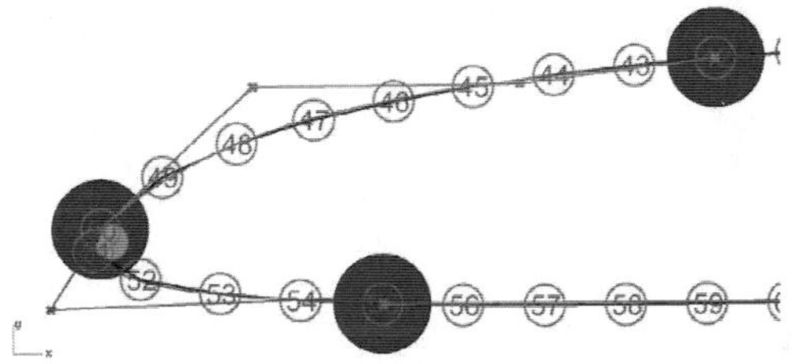

Figure 7: Detail of knot distribution at the leading edge, p = 3

4. Compute the knots values: once the B different Bézier pieces have been obtained with the previous steps, the knots k_i will be obtained as sum of lengths, divided by the total length:
5.

$$k_i = \frac{\sum_{1}^{i} L_j}{\sum_{1}^{B} L_j} \qquad i = 1, B-1 \tag{7}$$

Where L_j is the length of the Bezier piece of index j, and B is the number of pieces that models the profile data points. In the case of Fig. 6 B = 4, and the 3 knots will have the parametric values of 0.433, 0.516, and 0.556. If a uniform knot vector is employed, the parametric values would have been 0.25, 0.5 and 0.75.

Increasing the degree, produces more control points in the curves and this has the effect of diminishing the number of pieces that are necessary to model the data points. Fig. 8 shows the knots for a unit chord profile with 200 data points, fitted with p = 5, tol = 0.001 m. This section with a concave trailing edge can be found in marine hydrofoils. Notice how the leading edge is modelled with a single piece, and in this case the trailing edge concave part will need another one. In this case p = 5 means 6 control points per curve and B = 3, with the knots placed at 0.484 and 0.929. A uniform knot distribution would produce the knots at 0.333 and 0.666.

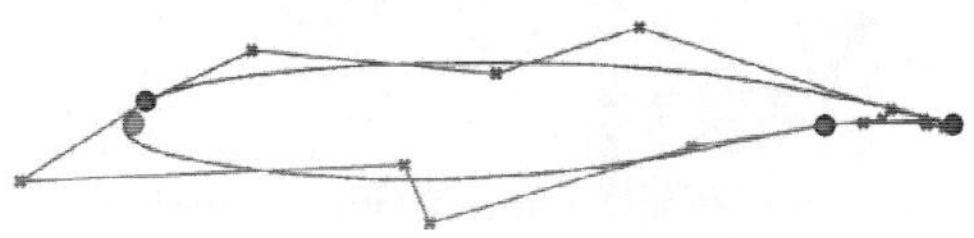

Figure 8: Example of knots distribution, p=5

It can be seen how most of profiles can be modelled with a little number of Bézier pieces, and that the leading edge will need one or two pieces to be reproduced accurately.

5. SUBDIVISION OF THE KNOT VECTOR CLOSE TO THE LEADING EDGE

Up to now, the position of the knots has been calculated according the previous sections. This determines the mathematical knot vector that is necessary to compute a B-spline according to Eq. (8).

$$[\underbrace{0,...,0}_{p+1}, k_{i=1,B-1}, \underbrace{1,...,1}_{p+1}] \quad (8)$$

Because of the properties of B-spline curves, the size of the knot vector (noted as *knots*), the degree p and the number of control points NP are related as $knots = NP + p + 1$. So, the number of control points is calculated directly once the knot position has been obtained following Section 4. In the example of Fig. 6 where $B = 4$ pieces, $NP = 11 - 4 = 7$ control points.

Nevertheless, this number of control points and knot vector does not guarantee that the tolerance is fulfilled when an ILS technique of the np+1 profile data points is computed using Section 3, where the fitting of the example of Fig. 6 has been computed with the previously mentioned 7 control points. The approximation of the leading edge can be quite poor, and the maximum distances to the data points is larger than the tolerance (red circles of 0.001 m).

This is solved in the present method by subdividing some Bézier pieces in a desired number P of equal parts, starting from the second piece, and finishing in the next to the last one. As depicted in the previous examples of Fig. 6 and Fig. 8, the leading edge is covered starting from the second piece until the next to the last one.

That means that extra knots are added in the knot vector close to the leading edge, and this way new control points will appear near this area in order to obtain a better fitting there. This effect is shown in Fig. 9 where the same example of Fig. 6 is fitted subdividing the 2nd and 3rd pieces in two. The knots (new mid-knots underlined) are now 0.433, <u>0.474</u>, 0.516, <u>0.536</u> and 0.556 in the parametric space, and the ILS technique of section 3 can be computed with 9 control points instead of 7. These two new control points associated with the mid-knots placed near the leading edge, have a great effect over the fore area that now matches the tolerance of 0.001 m.

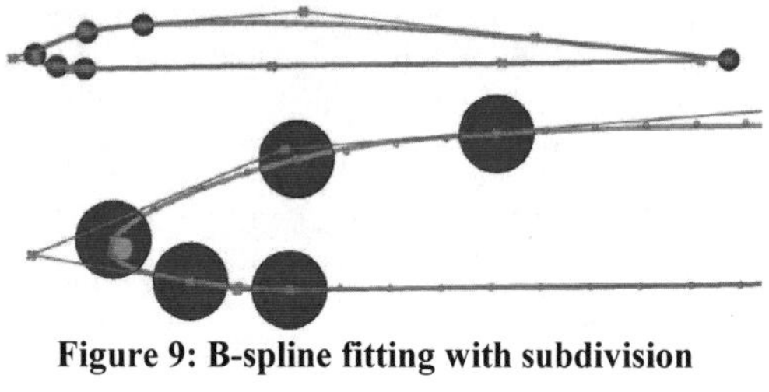

Figure 9: B-spline fitting with subdivision

Subdivision in 2, 3 and up to 4 parts according to the mentioned scheme has produced good results in the fitting of all the propeller profiles that the authors have tested.

6. FINAL ADJUSTMENT OF THE PROFILES

Some final improvements for the ILS method of Section 3 have been developed in this work. The reason is that Eq. (4) together with the way the parameters of the data points are obtained by iteration, can be unstable and produce strange results, especially if the number of data points is large and with high degree curves.

If one wants to place the control points of a 2D B-spline manually to resemble the profile data points, the control points will mimic the shape of the curve, and this cannot be the case when using the LS technique because of the afore mentioned reasons.

The improvement of the ILS presented in this paper is the addition of some kind of fairing of the control polygon. The problem is that the ILS only minimizes the distance between the points and the B-spline, and does not consider the shape of the resulting curve. This means that the result of Eq. (5) will be a curve that may squirm or curl producing big disturbances in the curvature, yet it is as close to the data points as a result of the ILS technique.

Fairing curves is a very subjective matter and there are plenty of criteria. In this paper a global fairing criteria is used using an integral formula, following (Piegl and Tiller, 1999) [11]. The fairing for a parametric curve *c(u)* will try to minimize function.

$$f = \int_0^1 \left\| \frac{d^2}{du^2} c(u) \right\| du \quad (9)$$

The problem arises because there is not a direct definition of the B-spline until the ILS method is applied and it is not possible to include Eq. (9) directly into the definition of the ILS fitting. The key is to change the focus from the curve itself to its control polygon, since it is inside the definition of the system (5) and is the final result of the ILS fitting. Because of the properties of the B-Splines, the curve mimics the shape of its control polygon: if the polygon conducts well (does not wiggle or twist), so does the B-Spline.

So, instead of applying Eq. (9) directly, the method considers the minimization of the second derivatives included in (9) changing derivatives by differences of the control points. If the differences of the curve's control points decrease, so do the derivatives of the B-spline. This can be accomplished setting the conditions of Eq. (10) to the control points for the second differences:

$$\begin{aligned} P_0 - 2 \cdot P_1 + P_2 &= 0 \\ P_1 - 2 \cdot P_2 + P_3 &= 0 \\ &\vdots \\ P_{N-2} - 2 \cdot P_{N-1} + P_N &= 0 \end{aligned} \quad (10)$$

This set of N-1 equations with N+1 unknown control points, can be developed in matrix form easily (11):

$$\begin{bmatrix} 1 & -2 & 1 & \cdots & 0 & 0 & 0 & 0 \\ 0 & 1 & -2 & \cdots & 0 & 0 & 0 & 0 \\ \vdots & \vdots & \vdots & & & & & \vdots \\ 0 & 0 & 0 & \cdots & 0 & 1 & -2 & 1 \end{bmatrix} \cdot \begin{bmatrix} P_0 \\ P_1 \\ \vdots \\ P_N \end{bmatrix} = \begin{bmatrix} 0 \\ 0 \\ \vdots \\ 0 \end{bmatrix} \Rightarrow [S] \cdot [P] = [0] \qquad (11)$$

So, the LS system of equations of Eq. (5) can be completed to include the minimization of the second derivatives, producing an extended matrix M_E of dimensions $(np + N) \cdot (N+1)$ and also an extended independent term Q_E with the form of Eq. (12).

$$\begin{bmatrix} B_0^p(t_0) & B_1^p(t_0) & B_2^p(t_0) & \cdots & B_{N-3}^p(t_0) & B_{N-2}^p(t_0) & B_{N-1}^p(t_0) & B_N^p(t_0) \\ B_0^p(t_1) & B_1^p(t_1) & B_2^p(t_1) & \cdots & B_{N-3}^p(t_1) & B_{N-2}^p(t_1) & B_{N-1}^p(t_1) & B_N^p(t_1) \\ \vdots & \vdots & \vdots & \ddots & \vdots & \vdots & \vdots & \vdots \\ B_0^p(t_{np}) & B_1^p(t_{np}) & B_2^p(t_{np}) & \cdots & B_1^p(t_{np}) & B_{N-2}^p(t_{np}) & B_{N-1}^p(t_{np}) & B_N^p(t_{np}) \\ 1 & -2 & 1 & \cdots & 0 & 0 & 0 & 0 \\ 0 & 1 & -2 & \cdots & 0 & 0 & 0 & 0 \\ \vdots & \vdots & \vdots & \ddots & \vdots & \vdots & \vdots & \vdots \\ 0 & 0 & 0 & \cdots & 0 & 1 & -2 & 1 \end{bmatrix} \bullet \begin{bmatrix} P_0 \\ P_1 \\ \vdots \\ P_N \end{bmatrix} = \begin{bmatrix} Q_0 \\ Q_1 \\ \vdots \\ Q_{np} \\ 0 \\ 0 \\ \vdots \\ 0 \end{bmatrix} \Rightarrow \qquad (12)$$

$$\Rightarrow [M_E] \cdot [P] = [Q_E]$$

To summarize the whole method, matrix formulation (11) contains an implicit fairing of the control polygon, and is included in the definition of the ILS fitting in Eq. (12) since these expressions are a function of the control points. A weighted combination of the derivatives/differences is used, and factor α inside the interval [0,1] that counts the effect of the differences will also be an input to ILS fitting. A large α produces a softer distribution of the control points, but may reduce the effect of fitting, increasing the distances between the data points and the B-spline. This system is shown in (13) and is solved as explained in Section 3.

$$\begin{bmatrix} (1-\alpha)\cdot[M] \\ \alpha\cdot[S] \end{bmatrix} \cdot [P] = (1-\alpha)\cdot[Q_E] \qquad (13)$$

The parameters t_i corresponding to the data points that are used in Eq. (12) and (13) are iteratively calculated based on the minimum distance from the data points to the B-spline, as explained in Section 3. Therefore, in order to obtain the fitting of the data points with a desired tolerance following the previous sections, the user has only to indicate the degree p, the number of parts P to subdivide the Bézier pieces that gives the non-uniform knot vector, and the value of α. Once that the profiles have been adjusted with B-splines, the surface definition can be computed.

7. LOFTING SURFACE OF THE PROFILES

The last step of the method is the definition of a B-spline surface that contain the previously defined B-splines. The generalization from curves to surfaces is not difficult due to the properties of B-splines, and thus, a lofting surface of the station pieces can be easily defined. In (Qian and Dutta, 2001) [6], some practical advices when lofting the propeller sections are presented in order to get better results at the propeller tip: they proposed the use of different types of final sections at $r=R$ considering a 5 to 1 chord / thickness ratio, so curvature in this area is improved. The tip is also improved if sections 0.975 r/R and 0.9875 r/R are included in the lofting.

The transition from spline curves to spline surfaces is achieved by turning the control polygon into a control net of control points $\mathbf{W_{ij}}(X_{ij}, Y_{ij}, Z_{ij})$ using the same B-spline basis for the two parameters u and v, as well as using two different lists of knots $\{u_{-1},\ldots,u_{N+n}\}$, $\{v_{-1},\ldots,v_{M+m}\}$, with the first list of knots obtained as described in the previous sections.

The lofting process of a set of q B-splines (profiles) with the same degree and knots is as follows. Find a B-spline surface **S** with degree n by m and $(N+1)$ by $(M+1)$ control points and knots $\{u_{-1},\ldots,u_{N+n}\}$ and $\{v_{-1},\ldots,v_{M+m}\}$ according to Eq. (14) that interpolates q different B-splines (3D Propeller profiles) noted as $\mathbf{c_d}$ $(d=0,\ldots,q-1)$ of nth degree with $N+1$ control points and a list of knots $\{u_{-1},\ldots,u_{N+n}\}$ with the form of Eq. (15).

$$\mathbf{S}(u,v) = \sum_{i=0}^{N}\sum_{j=0}^{M} \mathbf{W_{ij}} \cdot B_i^n(u) \cdot B_j^m(v) \qquad (14)$$

$$\mathbf{c_d}(u) = \sum_{i=0}^{N} \mathbf{V_{id}} \cdot B_i^n(u) \quad (d=0,\ldots,q-1) \qquad (15)$$

Note that $\mathbf{V_{id}}$ are the control points of the different profiles obtained in Section 3 and are expressed in matrix form. The values of N, M and q depend on user preferences regarding the definition of the blade surface. The interpolation can be written as:

$$\mathbf{S}(u,v_d) = \sum_{i=0}^{N}\left(\sum_{j=0}^{M} \mathbf{W_{ij}} \cdot B_j^m(v_d)\right) \cdot B_i^n(u) = \sum_{i=0}^{N} \mathbf{V_{id}} \cdot B_i^n(u) = \mathbf{c_d}(u) \quad (d=0,\ldots,q-1) \qquad (16)$$

This group of equations must be solved for a set of values of parameter v_d $(d=0,\ldots,q-1)$, which is called the choice of the parameterisation. Centripetal parameterisation produces good results for the shape of propeller blades. By identifying equal coefficients for each row in Eq.(16), $i = 0,\ldots N$, the following linear system is obtained:

$$\sum_{j=0}^{M} \mathbf{W_{ij}} \cdot B_j^m(v_d) = \mathbf{V_{id}} \quad (d=0,\ldots,q-1) \qquad (17)$$

To obtain a unique solution for this system, $M+1 = q$, where q is the number of profiles that define the blade. The $(M+1)$ by $(N+1)$ solutions are the control points $\mathbf{W_{ij}}$ of the lofting surface of Eq. (14) containing the profile curves.

8. MARINE PROFILES

In this section, different cases of 2D propeller sections of one meter chord are tested with the proposed method, obtaining maximum distances less than 1mm with a low number of control points. The earliest screw propellers were flat plates twisted into some pitch angle, producing a minimum lift based on the section shape. They were followed by ogival profiles with a flat face and a back circular arc. Gawn-Burril propellers (Dierckx, 1993) [12] used ogival shape into its definition and are still used today. With its flat face, the design is relatively easy to manufacture and repair, and this flat face is a common characteristic of several marine propeller profiles.

Further studies developed more efficient shapes such as the Troost profiles, or B-series propellers. These profiles have its maximum thickness somewhat forward of mid-chord and have a flat face. The series was extensively tested and used as the basis of numerical sizing methods. The fitting method is applied to a Troost profile of one meter chord and 210 points. In the example of Fig. 10, the degree p = 4, the fairing parameter α = 0.15 and the subdivisions of the leading edge P = 4. A tolerance of 0.001 m has been considered. Using these input values, the method produces a B-spline with 10 control points, that fits the 210 data with max / med distances of 0.766 / 0.248 mm. Computational time is about 3.0 sec. in a Intel Core I7 Processor, 3.4 GHz.

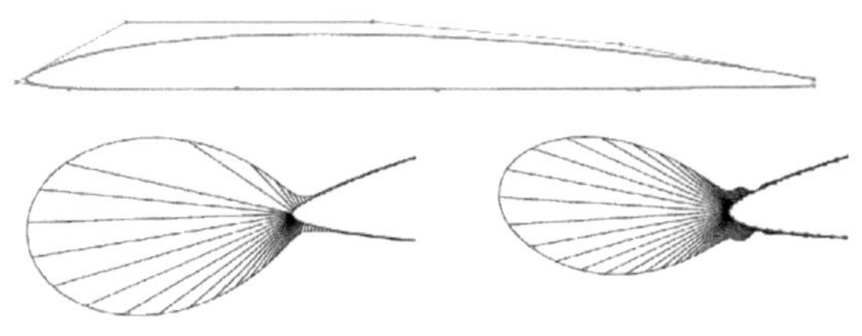

Figure 10: Troost profile, p =4, tol = 1mm (10 CP)

If an interpolation B-spline is used, 210 data points are obtained. The use of interpolating B-splines has also a background: the curvature graph may oscillate. In the curvature graphs of Fig. 10, the one on the left correspond to the B-spline obtained with the proposed method, while the one on the left is obtained with and interpolating spline. The graphical is not symmetrical in the last case in respect to the mean line, and this may have an influence in the lift and drag characteristics.

The approximation results can be improved by reducing the tolerance that will increase the number of Bezier pieces that are needed (Section 3) with the consequent increment of the knot vector size (Section 4) and hence the number of control points.

In the case of Fig. 11, the tolerance is set to 0.4 mm, p = 5, α = 0.2, And P = 4. The method produced a B-spline of 19 control points, with max/med distances of 0.389/0.116 mm that improves the approximation of Fig. 10. Computational time is about 3.4 sec. in an Intel Core I7 Processor, 3.4 GHz.

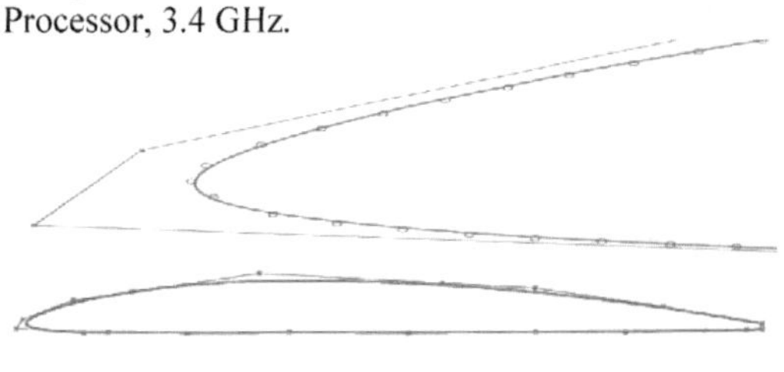

Figure 11: Troost profile, p = 5, tol = 0.4 mm (19 CP)

In this figure, a detail of the leading edge can be seen. Notice that the radius of the points is the tolerance 0.4 mm and the curve crossed all the data circles.

The research into blade section shape found that the B-series shape was not optimum, because of its pressure distribution that peaks strongly near the leading edge. Naval designers then began to use aircraft foil shapes and new designs were developed (Nowacki and Lü, 1994) [13] based on the NACA series. The shape no longer has a flat face which is common in today propellers although this curved face complicates repair.

Fig. 12 shows a one meter chord NACA 64212. These series are common in today propeller designs. The thicker leading edge and curved face, need more control points to be well reproduced. In this case, tol = 0.6 mm, p = 5, α = 0.6 and P = 2 produces a B-spline with 27 control points, that has a max/med distances of 0.518/0.133 mm. Computational time is about 3.6 sec. in an Intel Core I7 Processor, 3.4 GHz.

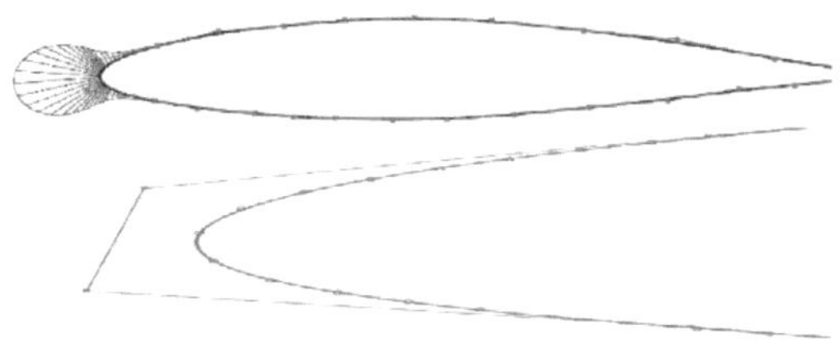

Figure 12: Modified NACA 64212, p= 5, tol = 0.6 mm (27 CP)

The research into blade section shape considered not only the leading edge, but also the trailing edge, especially in small craft propellers, than can present a propeller cup into the sections (Gawn and Burrill, 1957) [14]. A propeller cup is a deformation the its trailing edge toward the pressure face, allowing the blade to operate at a lower pitch with a lower suction peak to improve cavitation performance, while maintaining a flat face in the rest of the section.

Finally, the evolution of blade sections has bring out local modifications of the leading edge, or an increment of the thickness at the leading edge with a fuller nose (leading edge radius). This way the propeller will tolerate better variable angles of attack reducing the effect of cavitation. Example presented in Fig. 8, correspond to sections that can be used in marine hydrofoils. So, the presented method can reproduce well the profiles normally used in marine applications.

When modelling the leading edge of the propeller (Fig. 3), it must be accurately reproduced because most of the performance characteristics are induced by this region. This part of the surface presents high curvature values when compared with the rest of the shape of the blades and if it is modelled with a wrong definition of the leading edge, additional drag to the one that has been predicted will be present in the final design. The effect of the leading edge of a profile was investigated in (Scherer and Stairs, 1994) [15] applied to keel and rudder design.

9. A PROPELLER BLADE

A propeller blade modelled with the proposed method is presented in this section. It corresponds to a 5 bladed marine propeller. The purpose for the marine propeller model was to perform CFD calculations of the design, so the profiles were fitted with a degree $p = 5$, $\alpha = 0.55$ and a tolerance of 0.002 m trying to minimize the number of control points. The propeller diameter was 2.15 m and was composed of Troost sections. This tolerance was achieved subdividing the pieces in $P = 3$ parts as explained in section 5, obtaining B-spline with 11 control points. The presented method was also used to create propeller models, which were constructed with a propeller milling machine.

A 3D printed model of one of the blades was constructed directly based on the information generated in the presented method.

10. CONCLUSIONS

This paper has presented a practical method for producing a B-spline representation of propeller blades, based on its profile sections that are composed of a large number of data points, and also considering traditional blade design parameters and methodologies. The method first produces a fitting of the 3D profile data points allowing the leading edge, which has a major effect on the propeller properties, to be well reproduced and with a reduced number of control points, improving interpolating methods that produce a large number of control points.

The fitting also uses an iterative selection of the parameterization that permits setting a given tolerance, which is very useful from a construction perspective. The method produced propellers with a minimum number of control points that fits the data points below a given allowance. This reduced number of control points permits a simpler surface definition, which is important in subsequent stages of the design process including numerical CFD calculations and optimization. The method has been tested in Section 8 with different marine profiles used in propellers and hydrofoils, modelling well the leading edge for thin and thick sections, and also the trailing edge in the case of cupped propellers.

When compared with the manual process of designing a blade, the modifications can be made quickly and easily considering parameters that are familiar for the designer while producing a precise and exchangeable B-spline surface model.

The main originalities of this paper has been the use of the a non-uniform knot vector that depends on the shape of the propeller sections, as described in section 4, together with the use of the iterative parameterization depicted in section 3. With the use of these two key points, the main results of this paper has been a practical method for creating the surface of propeller blades according to parameters related with its performance, and subsequently modifying its design. Notice that this method can be easily adapted to rudders, keels or foils just considering the profiles into parallel planes instead of cylinders.

Most CAD systems are designed for general purpose surface modelling and allow shape change through direct geometric manipulation (control points). Since shape change based on blade parameters (pitch, rake, skew...) is not generally supported in commercial CAD systems, changing blade geometry can be a difficult and time consuming process. Propellers do not conform to linear XYZ space that CAD softwares use. They rotate and advance along an axis, so their natural coordinate system is helical. Considering these premises, the authors have developed this new methodology for a propeller representation.

REFERENCES

[1] Perez-Arribas, F. and Perez, R. *A B-spline Design Model for Propeller Blades*. Advances in Engineering Software. April. Volume 118. pp. 35-44. 2018

[2] Carlton, JS. *Marine propellers and propulsion*. 2nd ed. Elsevier Butterworth-Heinemann. 2007

[3] Adkins, CN. *Optimum design of propellers*. Journal of Propulsion and Power, **10/5**, 676-682. 1994

[4] Ye, X.; Jackson TR. and Patrikalakis NM. *Geometric design of functional surfaces*. Computer Aided Design, **28**, 741–52. 1996

[5] Hoschek, J. and Müller, R. *Turbine blade design by lofted B-spline surfaces*. J. Comput. Applied Mathematics. **119**, 235–248. 1999

[6] Qian, X. and Dutta, D. *Design of heterogeneous turbine blade*. Computer Aided Design **35**, 319–329. 2001

[7] Koini GN.; Sarakinos, SS. and Nikolos, IK. *A software tool for parametric design of turbomachinery blades*. Advances in Engineering Software. **40/1**. 41–51. 2009

[8] Kaplan, AR. and Macpherson, D. *Buildable propeller geometries: Answering the unique challenges for design and manufacture*. Proceedings of 13th Int. Conf. Computer and IT COMPIT14, pp.387-396. 2014

[9] Piegl, LA. and Tiller, W. *The NURBS Book*. Ed. Springer, p. 410-413. 1997

[10] Woodward C. *Cross-sectional design of B-spline surfaces*. Comput. Graphics **11**, p.193–201. 1987

[11] Piegl, LA and Tiller W. *Cross-sectional design with boundary constraints*. Eng. Comput. pp. 171–80. 1999

[12] Dierckx, P. *Curve and surface fitting with splines*. Oxford University Press. 1993

[13] Nowacki, H and Lü, X. *Fairing composite polynomial curves with constraints*. Computer-Aided Geometric Design, **11**, 1-15. 1994

[14] Gawn, RWL. and Burrill, LC. *The effect of cavitation on a series of 16 in model propellers*. Transactions of the Royal Institute of Naval Architects, **99**. 1957

[15] Scherer, O. and Stairs, R. *Propeller Blade Sections with Improved Cavitation Performance*. Proceedings of the SNAME Propellers/Shafting Symposium. 1994

[16] MacPherson, DM. *Small Propeller Cup: a Proposed Geometry Standard and a New Performance*

Model. Proceedings of the Propellers/Shafting '97 Symposium Virginia Beach. 1997
[17] Marchaj, CA. *Sail performance: theory and practice*. Ed. Adlard Coles Nautical. 1996

AUTHORS BIOGRAPHY

Rodrigo Perez Fernandez is the head of Naval Shipbuilding Business Development in SENER. He is also professor in several subjects at the Marine Engineering School of the Technical University of Madrid.

Francisco Perez-Arribas he is also professor in several subjects at the Marine Engineering School of the Technical University of Madrid.

THE IMPORTANCE OF A NON-DETERMINISTIC DESIGN OPTIMIZATION FOR PREDICTING REAL-LIFE PROPELLER PERFORMANCES

K. Vidal, B. Mallol and **C. Hirsch,** NUMECA International, BE
L. Poppelier, SIP Marine BV, NL

SUMMARY

The manufacturing of propellers comes with geometrical variability that lies within predefined tolerances decreed by standardized accuracy classes. Although controlled by these tolerances classes, the variation in the design might result in a degradation of the expected propeller behaviour.
In a classical deterministic design optimization, engineers improve a digital model, without taking into account these variations. Hence a new type of optimization method is recommended whereby the impact of the manufacturing variability, represented by statistical moments, is minimized.
This paper presents a robust non-deterministic optimization of a ducted marine propeller mounted on an inland vessel. In this test case, the statistical moments of the propeller efficiency are optimized while axial thrust is constrained and cavitation occurrence is considered. The manufacturing uncertainties are derived from the ISO geometrical tolerances S-class. Eventually, a robust optimum is compared with a deterministic optimum in order to underline the benefits of the non-deterministic design methodology.

NOMENCLATURE

Upper case symbols

D	Propeller diameter (m)
J	Advanced ratio (-)
Kt	Thrust coefficient (-)
Kq	Torque coefficient (-)
P	Pressure (N m^{-2})
Pvap	Vapor pressure (here 1700 Pa)
Q	Torque (N m)
T	Axial thrust (N)
V	Far field velocity (m s^{-2})

Lower case symbols

n	Revolution rate (rps)

Greek letter

η	Open water efficiency (-)
μ	Statistical mean (units of the quantity)
υ	Kinematic viscosity (N s m^{-2})
ρ	Density of water (kg m^{-3})
σ	Standard deviation (units of the quantity)

Abbreviations

AIS	Automatic identification system
BE	Breath extreme
CAD	Computer aided design
CDF	Cumulative density function
CFD	Computational fluid dynamics
CPU	Central processing unit
DoE	Design of experiment
GA	Genetic algorithm
ISO	International standardization
LHS	Latin hypercube sampling
LOA	Overall length
NIPColM	Non-intrusive probabilistic collocation Method
PDF	Probability density function
RDO	Robust design optimization
RPM	Rotation per minute
SA	Spalart–Allmaras
SBDO	Simulation-based design optimization
UQ	Uncertainty quantification

1. INTRODUCTION

In the recent year, the design time has been drastically reduced thanks to modern computing involving computational fluid dynamics (*CFD*) and embedded and automated frameworks for the simulation-based design optimization (*SBDO*). Nowadays, hydrodynamics *SBDO* of a marine propeller is currently run with a unique set of inputs, such as geometry parameters and operating conditions. This overwhelming deterministic approach produces a single value response, that mainly drives all the optimization strategies. But the actual operating conditions are subject to uncertainties coming from manufacturing tolerances, operating conditions and long term life cycle involving the slow degradation of materials because of cavitation or fouling.
Aware of these variabilities, modern designers currently use tools that generally do not provide means to assess the impact of these variabilities. And consequently, they need to impose very strict and costly manufacturing tolerances to make sure the newly optimum design is not influenced.
This paper presents a non-deterministic methodology that allows to assess quantitatively the effect of these variabilities and is applied on the design optimization of a ducted marine propeller. This newly optimization procedure, called *RDO*, aims at improving the performances but also guaranteeing a stable behaviour given an inputted variability.
Hence, the *RDO* provides a range of confidence with the simulation results and might help future designers in their design strategy.

In this work, two operating conditions are studied, and the manufacturing tolerances come from the ISO 484-2 2015.

2. PRESENTATION OF THE TEST CASE

2.1 BACKGROUND

The background of the project consists in optimizing a propeller mounted on an inland vessel with a single propeller system and twin-blade rudder appendages.

Figure 1: The inland vessel under different loadings.

The inland vessel market will grow at over 4.5 per cent from 2017 to 2024 [1], justifying the recent trend at improving the performances of such vessel for economic and ecological reasons [2]. Consequently, the optimization of the propulsion device could be an achievable solution for naval architects.

AIS vessel type	Tanker - Haz C
LOA x *BE*	109.7 m x 11.4 m
Deadweight	2810 tons
Year built	2010
Design draught	2.8 m
Max draught	3.35 m
Design speed	5.4 knots (10km/h)
Max speed	6.2 knots (11.5km/h)

Table 1: the inland vessel main particulars.

2.2 PRESENTATION OF THE DUCTED MARINE PROPELLER

The propeller studied is a fixed pitch ducted propeller where the main characteristics are defined in *table 2*.

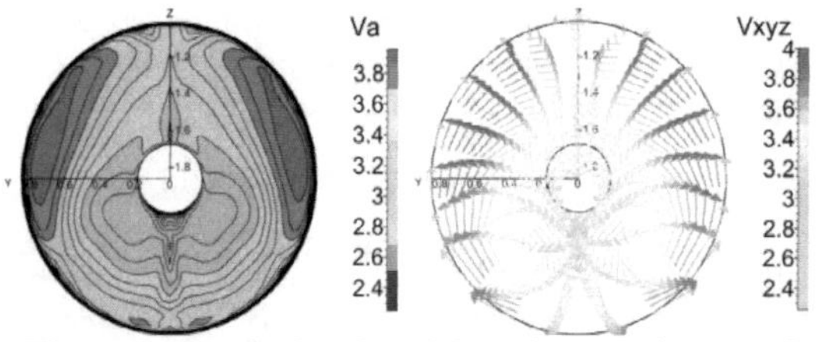

Figure 2: Nominal wake of the ship at design speed.

According to [3] all newly designed inland vessels have propellers with a nozzle. The main reason is the capacity to carry on higher thrust loading with better efficiency.

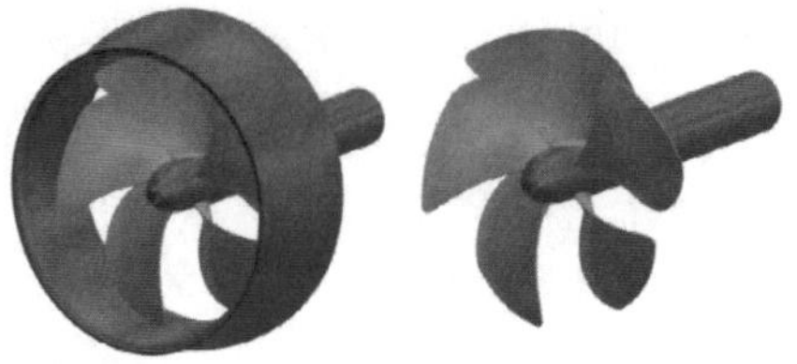

Figure 3: The ducted propeller with and without the nozzle.

The studied case is a replacement propeller for the vessel. No design specifications were formulated for the nozzle, because it is embedded in the hull of the ship (cf. *fig.4*).

Figure 4: Final propeller mounted on the ship, the duct is part of the hull

In general practice, this type of propellers is based upon the Wageningen Ka-series ducted propellers. Hence, the empirical formulas, based on $J\text{-}K_t/K_q$, determine thrust and power as a function of speeds and engine characteristics. The working area of the propeller is eventually defined, and therefore the blade area ratio and the required pitch.

Type	Fixed pitch
Diameter	1.7 m (2.08 m with duct)
Number of blades	5
Mean pitch	2.251 m
Chord length at 0.7R	0.68 m
Revolutions/sec	5.385 rps

Table 2: the propeller main particulars.

Then, the ship resistance is computed in order to extract the wake field.
This field helps to finalize the design loop by adjusting the section profile, and more shape-determining parameters. Hence the optimization procedure can be applied during those final iterations in the design process.

3. DETERMINISTIC OPTIMIZATION FRAMEWORK

3.1 GENERAL PRINCIPLE

The most straightforward optimization approach is to sample the design space and use *CFD* at the selected point and try to find a design for which the performances are better. However, each optimization iteration requires a *CFD* computation, increasing the time and computation power. In this manner, an efficient methodology is implemented in

FINE™/Design3D, and relies on two building blocks, pre-exploration of the design space, performed by a *DoE*, and a surrogate assisted optimization.

3.2 DESIGN OF EXPERIMENT FOR DESIGN SPACE EXPLORATION

The first step consists in pre-exploring the design space, by using a near-random *DoE*. This pre-exploration allows creating a metamodel or a response surface that is going to be used during the optimization. For that purpose a *LHS* is applied to sample the design space.

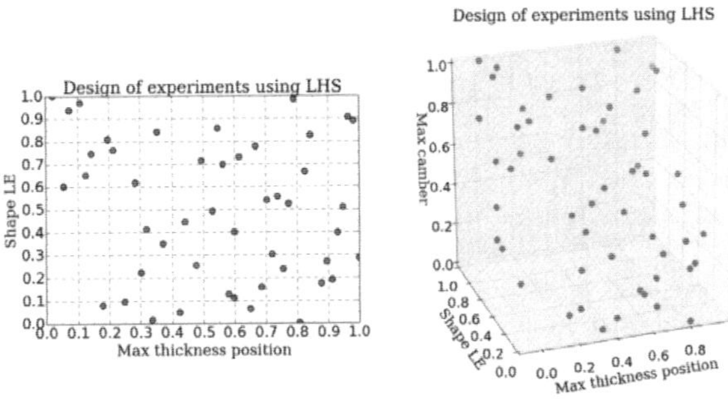

Figure 5: *DoE* of 30 samples for two and three design parameters.

As the response surface accuracy depends on the *DoE*, the number of initial design to generate is a compromise between:

- a high number of samples creating an accurate approximate model, but refining in area of low interest and requiring more computation power;
- a low number to limit as much as possible the *CFD* effort.

For a full benefit of this pre-exploration, the *DoE* is populated with five times the number of design parameters, in this study.

3.3 SURROGATE MODEL

From the *DoE*, the metamodel can be created. This approximate model, also called surrogate model, provides global design trends. The aim of the surrogate model is to mimic the propeller performances at points in between the *DoE* samples with a significant lower cost than *CFD*.

3.4 DETERMINISTIC SURROGATE ASSISTED OPTIMIZATION

The optimizer uses a genetic algorithm (*GA*) to locate the interesting area. *GA* has the advantage to search globally in the design space [4] and mimics the natural evolution process by extracting the best design candidates over a population based on the surrogate results. Once the *GA* coupled with the surrogate model finds the expected optimum. A *CFD* simulation is performed in order to check the validity of this expected optimum design and fine-tuned the response surface for higher accurate prediction capacity with the aim of advancing further towards the global optimum. The objectives and constraints formulations can be either single objective. In case of several operating conditions, the objectives and constraints can be aggregated. Or multi-objectives, where the outcome of the optimization is the Pareto front. The Pareto front represents a set of non-dominated designs, meaning there is not design better for the respective objectives combination. The non-deterministic optimization, discussed in the next section, shares a common workflow with the deterministic one. The main difference lies on the definition of the objectives and constraints.

4. NON-DETERMINISTIC OPTIMIZATION FRAMEWORK

Any industrial design is influenced by a superposition of several uncertainties. In order to understand the impact of these uncertainties, an uncertainty quantification (*UQ*) method needs to be applied.

4.1 UNCERTAINTY QUANTIFICATION METHOD

4.1 (a) Brief technical overview of the methodology

To allow the propagation of uncertainties, the method, used within this work, is the non-intrusive probabilistic collocation method *(NIPColM)* [5].
This approach was implemented and successfully applied on the *NASA rotor 37* validation case in [6].
The basis of this method is formed by the expansion of the solution into Lagrange interpolating polynomials.
The base points are the collocation points, which correspond to the Gauss quadrature points weighted by an input uncertainty defined as probability density function (*PDF*).
In order to compute the Gauss quadrature, the Golub-Welsch algorithm is used to provide the collocation points and it weights [7].
A system of uncoupled simulations can be eventually derived. Once the uncoupled simulations are solved, statistical moments of any output φ_i are automatically computed by taking the weight ω_i from the Gauss quadrature. The mean and variance are calculated as followed:

- For non-centered moments, such as the mean for $n = 1$;

$$\mu_{n_{non-centered}} = \sum_{i=1}^{Np} \omega_i (\varphi_i)^n \quad (1)$$

- For centered moments, such as the variance ($n = 2$).

$$\mu_{n_{centered}} = \sum_{i=1}^{Np} \omega_i (\varphi_i - \mu_{1_{non-centered}} \quad (2)$$

These statistical moments express the measure of the uncertainties influence over the quantities of interest.

4.1 (b) Sparse grid quadrature for solving multiple uncertainties

In order to handle multiple uncertainties simultaneously, the *NIPColM* is usually applied with a full tensor product. This eventually leads to an exponential number of *CFD* simulations which is referred in the litterature by *"the curse of dimensionality"*, and therefore cannot be used at industrial scale.
To solve this issue [6] have applied the sparse grid technique, based on Smolyak's quadrature method [8]. The sparse grid technique consists in truncating dimensional function terms higher than a level of accuracy. Thus, Smolyak's quadrature is not subject to the curse of dimensionality, and therefore offers a more efficient way of approximating function in high dimension, allowing so to make simultaneous treatment of many uncertainties in complex *3D CFD* simulation.

Dimension	Tensor product grid	Smolyak grid Level=1
1	3	3
2	9	5
10	59049	21
20	3^{20}	41

Table 3: Tensor product grid with 3 points in each dimension versus Smolyak grids using a level 1.

4.1 (c) Sensitivity analysis for understanding the relative uncertainty influence

When facing multiple uncertainties, the analysis of the relative influence of each input uncertainty is an important element. This analysis is based on the scaled sensitivity derivatives, introduced by [9], and is used to identify the most important uncertainty over an output quantity.
In practice, the sensitivity derivative is defined as the partial derivative of the solution, from the system of uncoupled equations, with respect to the input uncertain parameter. The result is then scaled by multiplying the standard deviation of the same random input.
The scaled sensitivity derivatives provide so the influence of each parameter separately, so no combined effects are taken into account.
Moreover, by providing a measure of the influence of uncertainties over the output, this analysis allows to reduce the number of uncertain input, by removing input with little influence. This can have the positive effect of using less deterministic simulations.

4.2 ROBUST DESIGN OPTIMIZATION

The main difference between the deterministic and non-deterministic optimization is the definition of the objectives and the constraints. The prior considers single values, and the latter takes into account statistical moments, represented by the mean and the variance (*eq. 1* and *eq. 2*). The most direct approach is to apply the *UQ* method for every single design in the *DoE*, hence the surrogate model is constructed based on the statistical moment of each sample. However, the cost of the non-deterministic *DoE* corresponds to the cost of the deterministic one times the number of *UQ* simulations. For a usual hundreds of samples in the *DoE*, this approach is beyond any industrial application.
For this reason [10] proposes the solution to build the *DoE* including both design parameters and uncertainties.
Here is the example of 18 design variables and 4 uncertainties assuming the distribution is symmetric, it gives 9 *UQ* simulations to run. Using 3 points in each direction of the *DoE,* we can obtain the following:

- *DoE* with *UQ* simulation:
  $18 * 3 * 9 = 486$ *CFD* simulations
- Mixed *DoE*:
  $(18 + 4) * 3 = 66$ *CFD* simulations

This gain in the computational time is achieved at the expense of not having the statistical moments directly available in the surrogate model. In this manner, the surrogate is updated with *UQ* simulations during the optimization procedure.

5. PRE-OPTIMIZATION STUDY

5.1 NUMERICAL MODELLING

5.1 (a) Grid generation and grid dependency study

AutoGrid5™, used for the meshing, is an automated multi-block structured mesh generator, that gives a high quality mesh on the surface, for a short generation time, in our case a 5 million points mesh is generated in 2 minutes on 4 threads.
Moreover, templates can be used to project same topology grid into newly design geometry shape, which is a main advantage in the frame of an optimization.

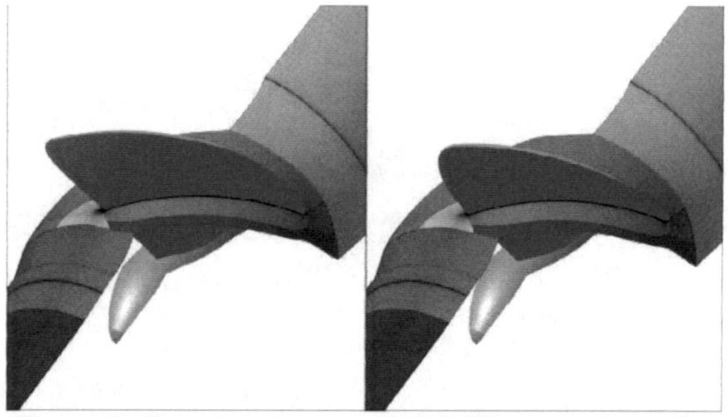

Figure 6: Template in AutoGrid5™ helps to keep the exact same mesh topology for different geometry shape.

A O4H topology is used to generate a serie of five nested meshes. The targeted *Y+* is 100, since the study is carried out on full scale. It gives a first layer size normal to the surface equal to *8e-05* meters. The O4H grid topology consists in a O-block in the skin block of the blade and H-blocks around the main skin block. The quality of the nested meshes are summarized below (cf. *table 4*). The meshes are obtained by adjusting the number of points in the O4H topology without modifying the first layer thickness. It shows that the meshing strategy has maintained the high quality mesh and the *Y+* requirements.

Mesh	1	**2**	3	4	5
Millions points	3.9	**5.0**	6.2	7.8	9.76
Min. skewness [°]	10.4	**10.4**	10.4	10.4	10.4
Max. exp. ratio [-]	2.8	**2.9**	2.6	2.3	2.0

Table 4: Overview of the quality of the five nested meshes.

At the end, it is found that a mesh of 5 million points provides an excellent trade-off between accuracy and computation time (cf. *fig. 8*).

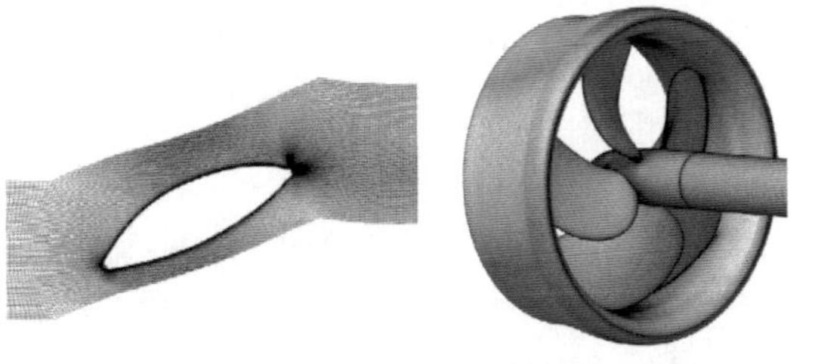

Figure 7: Overview of the final mesh, on the left, a blade to blade view at *0.2R*, and on the right, the mesh repetition.

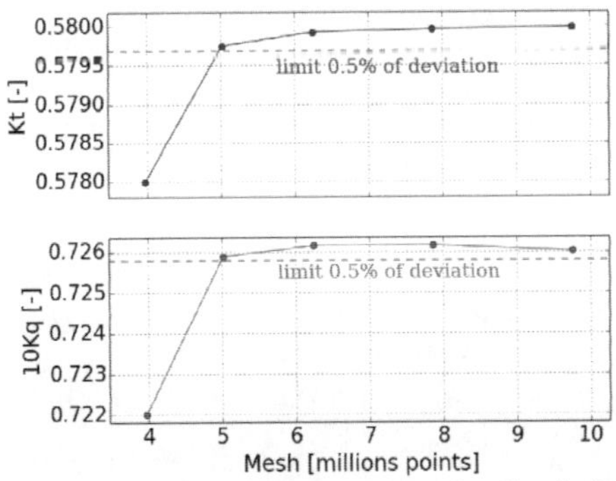

Figure 8: Grid convergence curve for the design operating condition.

5.1 (b) Physics and boundary conditions

FINE™/Turbo, used as flow solver, has been validated for open water computations, and shows excellent agreement with regards to experiments, and also to capture cavitation pattern [11].

FINE™/Turbo is a three-dimensional density-based structured multi-block Navier-Stokes solver using finite volume. Central-space discretization is employed along with multi-grid, local time-stepping and implicit residual smoothing, in order to speed-up the convergence. For the current propeller, the grid density allows to get a 4 levels grid and a V-cycle for the multi-grid method.

In case of low mach number incompressible fluid, time-marching density-based solver lacks efficiency, a low speed preconditioning is often required to improve the convergence rate. The preconditioning method presented in Merkle [12] is applied.

Standard boundary conditions for a typical open water computation are applied and defined below:

- A static velocity profile and a static turbulence conditions are specified at the inlet;
- An atmospheric pressure is applied at the outlet;
- A slip wall is set on the side wall of the cylindrical domain, which allows to monitor the internal mass flow for convergence behaviour;
- Periodic conditions are used to simulate only one blade passage;
- And the characteristics of the fluid are described in the *table 5* below.

Fluid model	Incompressible
ρ	998 kg/m^3
υ	1.0453e-06 m^2/s

Table 5: Main characteristics of the fluid.

The choice of the input velocity is dictated by the simulation of the ship at 10km/h and 11.5km/h. It eventually gives the velocity profile at the propeller location (cf. *fig.2*).

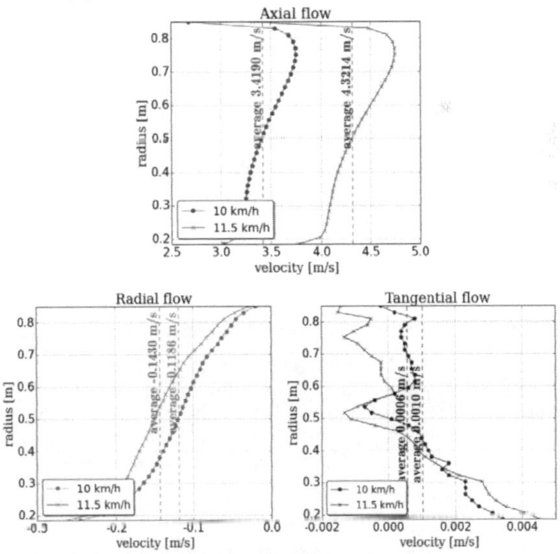

Figure 9: Circumferential averaging of the wake flow field.

In order to take into account, the highly non-uniform wake while assuming a steady state periodic computation, the wake is averaged in the circumferential direction over different radius locations (cf. *fig.9*).

It can be seen on *fig.9*, the radial component can be neglected, and the tangential one remains small. Hence only an axial profile is used at the inlet.

The position of the boundary is also important and requires considerations.

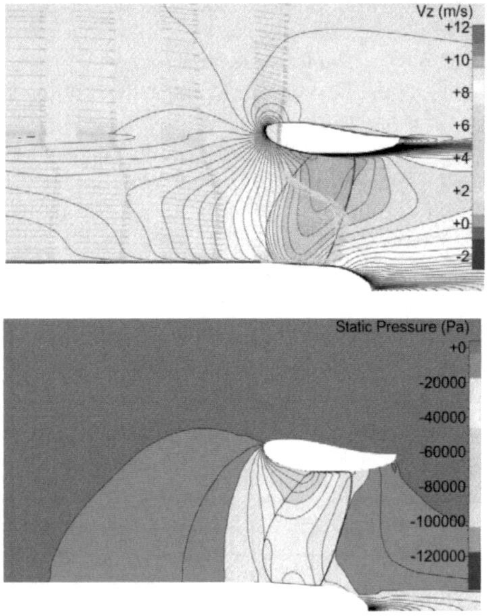

Figure 10: Meridional view of the simulation using the 11.5km/h velocity profile, with the axial velocity at the top, and the extrapolated pressure at the bottom.

The inlet is eventually placed 1 diameter length, for ensuring an interaction between the inlet condition and the blade, without creating an upstream extrapolated lower pressure level (cf. *fig.10*), and 5 diameter length for the outlet.

This configuration is used for both operating conditions (10km/h and 11.5km/h).

To simulate the three dimensional environment, Spalart–Allmaras model of turbulence is selected. At the end, the computation lasts 26 min on 48 cores.

5.1 (c) Cavitation inception

Cavitation can have a significant influence over the open water characteristics of the propeller, and thus should be considered when designing and optimizing any marine propeller. The complexity of the physics is real, and designers often use costly cavitation model, to evaluate its presence and the impact of this two phase phenomenon.

In the frame of the optimization, the usage of cavitation model is source of higher *CPU* cost and can be contradictory with the expected short design process. Hence a meaningful way to predict the cavitation is to compare the pressure level on the blade with the vapour pressure.

In our study, a *15°C* fresh water is assumed with a vapor pressure equal to *1700 Pa*.

In order to monitor the cavitation, the cavitating volume, or cavity, must be extracted. As the propeller tip position is quite close to free surface position at the design condition, the column of water is neglected. Hence the cavity is computing by mean of an iso-surface of static pressure below *Pvap*.

This method has proved to give a first idea of the cavitation pattern [11], but must be validated by a more accurate model afterwards. But in the case of the optimization, it is an excellent quantity to monitor.

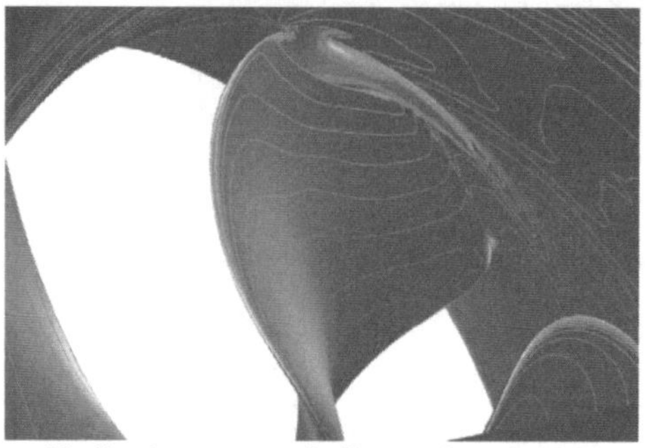

Figure 11: Static pressure isoline and cavitating cavity, represented by an iso-surface in blue, for the design condition (10 km/h).

The cavity of about *400 ml* remains small compared with the overall size of the propeller. Besides, [13] indicates that a moderate level of cavitation may not affect the propeller hydrodynamics performances, so the constraint on the cavitation volume should not be too strict during the optimization.

5.2 SELECTION OF THE UNCERTAIN PARAMETERS

5.2 (a) Geometrical uncertainties

Manufacturing tolerances, for marine propellers, are ruled by *ISO* norms such as the *ISO-484-2* for any marine propeller between *0.80* and *2.50m* [14].

The tolerances are usually expressed as a lower and a upper deviation from a nominal value. Those extrema are taken to represent a statistical tolerance by means of a probability density function (cf. *fig.12*).

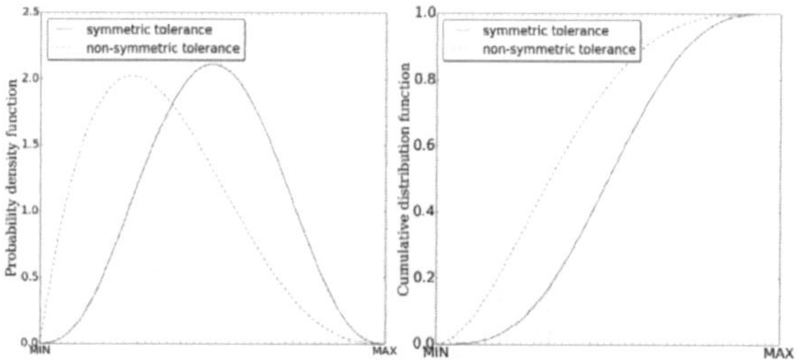

Figure 12: *PDF* and *CDF* for symmetric and non-symmetric distributions.

In this paper a beta-distribution is used and determined from the *ISO* extrema which is a percentage of the deviation compared to the nominal design value.

	Plus tolerance	Minus tolerance
Radius	0.2% (not less than 1.5mm)	0.2% (not less than 1.5mm)

Rake	0.5%	0.5%
Blade thickness	2% (not less than 2mm)	1% (not less than 1mm)
Blade chord	1.5% (not less than 7mm)	1.5% (not less than 7mm)

Table 6: Definition of each tolerance in case of a S-class manufacturing accuracy.

Those extrema are globally prescribed by four accuracy classes, ranging from the most restrictive to the widest tolerances. In this test case, the S-class tolerance is used, which represents the most accurate and restrictive one.
The tolerances definition is of importance, because the final parametric modeller of the propeller has to be defined taking into account the means of tolerance measurements.

5.2 (b) Operational uncertainty

Inland vessels operate seldom at the same working regime. The ship may be navigating at different speeds, especially speeds higher than the design point, or with different loadings inducing different draughts (cf. *fig.1*).
Those variations might change the ship wake, influencing the operating regime of the propeller.

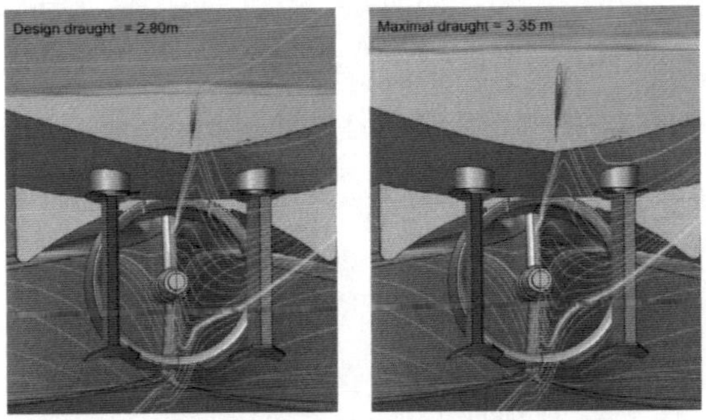

Figure 13: Streamlines, colored by the velocity magnitude, at the wake of the ship for different draughts.

In this manner, two resistance computations are performed with two different draughts, nominal and maximal draughts (cf. *tab. 1*). The nominal wake is then computed.
Fig.14 shows the difference in the velocity field at the propeller location. In order to enlighten the variation of the axial velocity, a radial averaging is performed, and shows a maximal variation of 3%. Hence, a symmetric *PDF* is defined using as extrema the maximal variation.
The wake axial profile (cf. *fig.9*) is going to be shifted by this variation during the *UQ* computations.

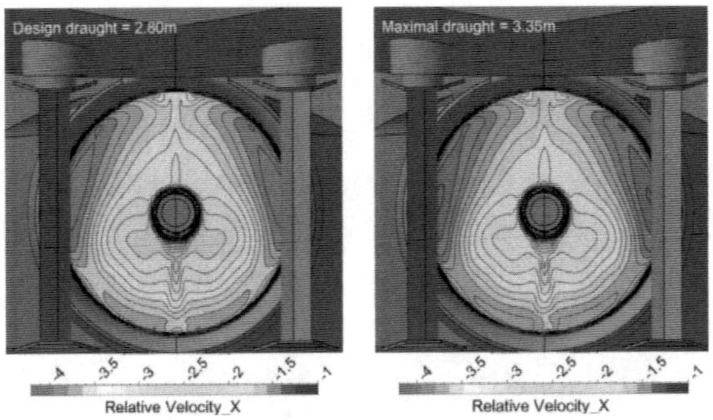

Figure 14: Relative axial velocity at the propeller location for the design speed condition under two different draughts.

5.3 PARAMETRIC PROPELLER DEFINITION

In order to construct the parametric model, the target is first sliced at different radius locations (cf. *fig. 15*). From those sections, a parametric topology is used to define the section profile and camber. The stacking of each section, in other words their relative position, defines the rake in the meridional direction and the skewness in the tangential one. Both meridional and tangential laws are parametrized using a Bezier of 6 control points.
The profile is defined using a B-spline curve of 6 control points and the camber using a Bezier of 3 control points. This Bezier curve also defines the local profile pitch and chord.

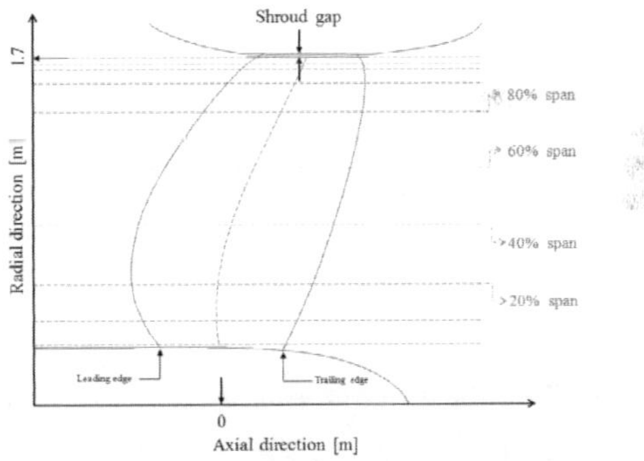

Figure 15: Location of the slicing sections along the span.

The choice of the span locations is determined by the *ISO*. Indeed, the *ISO* requests manufacturers to verify the deviation at different locations on the blade, i.e. 20%, 40%, 60% and 80% of the radius. Thus, it is better to control the uncertainty parameter and apply a correct evaluation of its impact when both manufacturing measurement and parametric modification are performed at the same radius.
As suggested by [10], the parametrization induces modifications and deviations from the initial target geometry. In case of uncertainty study, it is crucial to represent the initial target geometry with a high accuracy, such that the previously discussed deviations are significantly lower than the geometrical uncertainty.

This verification is performed on two levels:

- First the parametric and the target geometry are visually compared;
- Then the *CFD* simulations are juxtaposed.

5.3 (a) Accuracy of the geometry parameterization

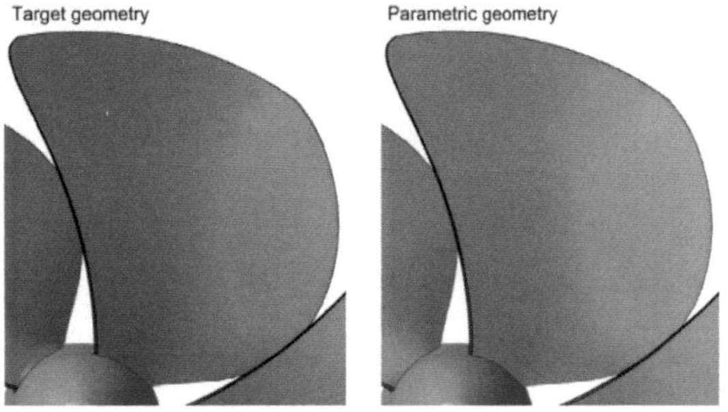

Figure 16: Comparison between the target (left) and the parametric geometry (right).

Figure 16 shows both geometries without displaying a significant difference. Hence a close-up is performed on different sections of the blade, and shows discrete geometrical deviations and amount locally up to 500 microns (cf. *fig.17*), where the uncertainty is equal to 2 mm for the thickness and 7 mm for the chord.
In conclusion, the local modifications are in overall lower than the manufacturing tolerances.

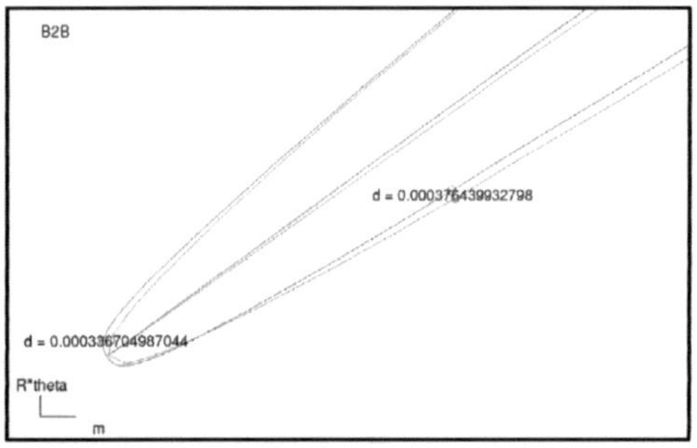

Figure 17: Close up at the leading edge part at 60% of the span.

5.3 (b) Influence of the parametric model on the deterministic results

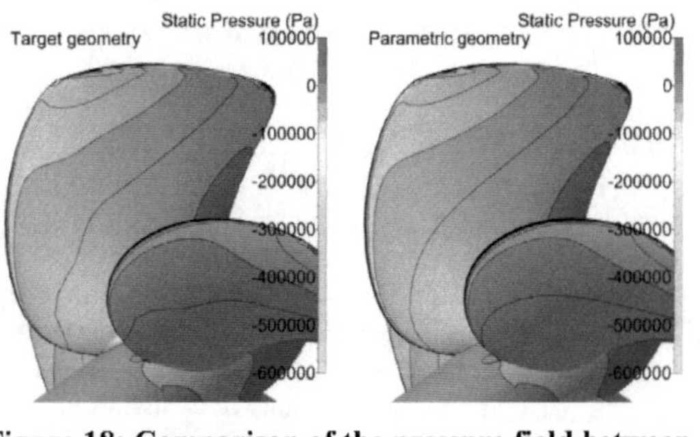

Figure 18: Comparison of the pressure field between the target (left) and the parametric geometry (right).

The influence of the error introduced by the parametric model on the *CFD* results must be assessed.
This also allows validating the parametric model as starting design for the optimization.
Eventually, the flow field does not show any serious difference, and the overall results are within 1%. It validates the parametric and also implicitly confirms that even a small variation in the geometry discretely affects the performances. It is important to underline that the evaluation of the parametric model accuracy has an importance in the *RDO* framework. For deterministic optimization, this procedure can be more flexible, since the geometry is deformed in any case.

	Target	Parametric	$\frac{Parametric - Target}{Target} \times 100$
$T\ [N]$	1.411e+05	1.402e+05	-0.71%
$Q\ [Nm]$	2.959e+04	2.985e+04	+0.87%
cavitating volume	400 ml	398 ml	-0.5%

Table 7: Comparison of the target and parametric geometry.

6. DETERMINISTIC AND NON-DETERMINISTIC OPTIMIZATION

6.1 DETERMINISTIC OPTIMIZATION

6.1 (a) Overall formulation

The design parameters consist in a total of 18 variables:

- 4 for the sections profile law;
- 2 for the chord law;
- 3 for the local pitch law;
- 3 for the camber law;
- 6 parameters for the skewness and the rake laws.

The thickness is kept constant, as the modification would induce a change on the structural computation of the bending moment. As suggested in section *3.2*, the *DoE* is sampled by 90 designs. For each design in the *DoE*, the following quantities are automatically post-processed, along with the cavitating volume:

$$K_t = \frac{T_{propeller} + T_{nozzle}}{\rho n^2 D^4} \quad (3)$$

$$K_q = \frac{Q_{propeller}}{\rho n^2 D^5} \quad (4)$$

$$\eta = \frac{K_t J(Va, D)}{K_q 2\pi} = \frac{K_t}{K_q 2\pi} \frac{V_a}{nD} \quad (5)$$

Two operating conditions are considered and linked to the ship speed (cf. *table 1*), 10 km/h and 11.5 km/h. The revolution rate of the propeller is kept constant. Thus, only the inlet flow condition is changed (cf. *fig.9*).

The overall objectives and constraints are the following:

- Improving the overall open water efficiency η by using an aggregated formulation of both conditions $0.7 \times design\ speed + 0.3 \times off\ design\ speed$;
- Maintaining the axial thrust as it is a requirement for the contractual ship design speed. So, K_t is constrained and equal to *0.58*;
- Not deteriorating the performance, by keeping a similar cavitating behaviour. As this constraint should be flexible (cf. *section 5.1(c)*), it is decided to keep the volume below *800 ml*;
- Simulation should be converged. The criteria, to decide whether the simulation is converged or not, is based on the difference between the inlet and outlet mass flow. This deviation should not exceed 0.1%, otherwise the sample is rejected.

6.1 (b) Optimization outcome

The deterministic mono-objective optimization reaches an optimum meeting all constraints within 30 iterations. The final optimum shows an increase in the weighted efficiency of about 2.26% relatively (i.e. almost 1% in absolute), while the thrust is maintained and the cavitation is below the fixed threshold.

The *table 7* summarizes the performances of the deterministic optimum compared to the baseline, the initial parametric geometry (cf. *fig.16*).

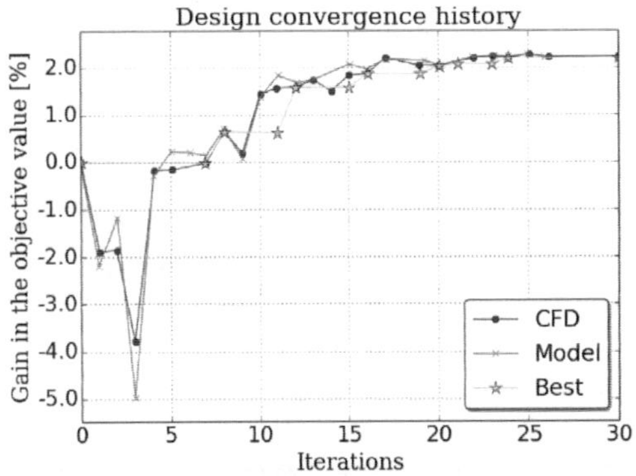

Figure 19: Convergence history of the mono-objective deterministic optimization.

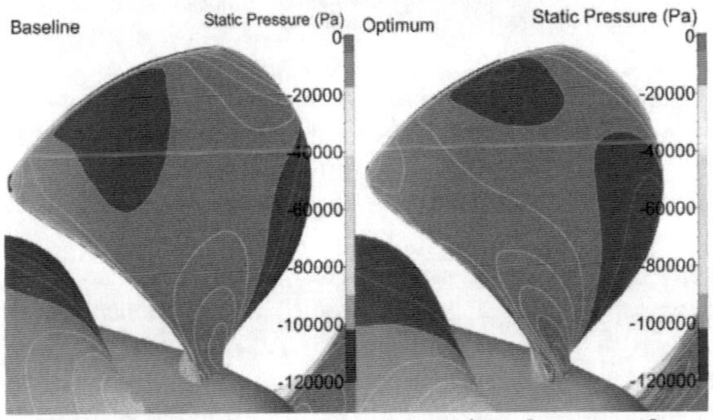

Figure 20: Static pressure comparison between the baseline (left) and the deterministic optimum (right).

When looking at the downstream velocity field, the tangential velocity of the optimum is less pronounced than the baseline (cf. *fig 21*). This indicates a more aligned flow and consequently a higher efficiency. This is also implied by the *figure 22*, where the midspan profile drag of the baseline is higher.

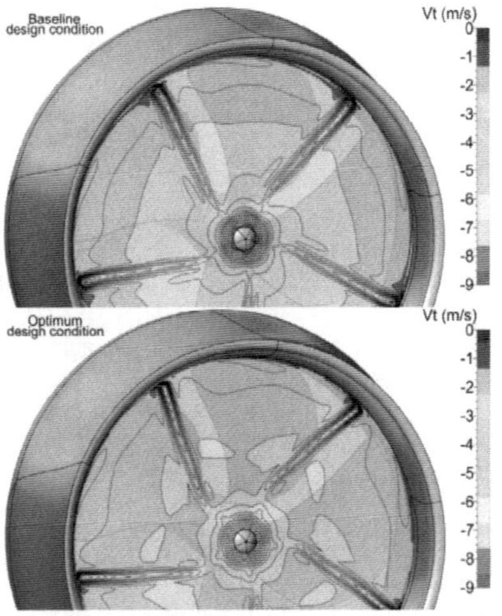

Figure 21: Tangential velocity behind the trailing edge for the baseline (upper picture) and the deterministic optimum (lower picture).

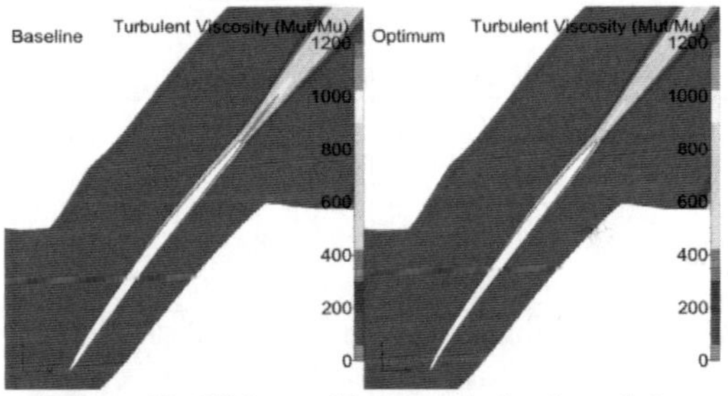

Figure 22: Midspan blade to blade view of the turbulent viscosity between the baseline (left) and the deterministic optimum (right).

For the cavitation modelling, it can be seen that the improvement in the propeller performances is highly constrained by the cavity volume (cf. *table 7*), as the final optimum features a cavity volume closed to the constraint value *800 ml*. In other words, the optimizer can not find a better optimum without exceeding this constraint.

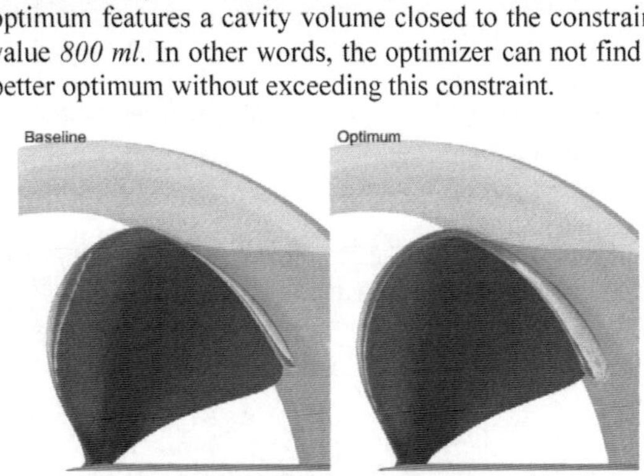

Figure 23: Cavitating simulation for the baseline (left) and the optimum (right).

Two steady cavitating simulations are eventually run to validate the actual performances of the propeller (cf. *fig.23*) and demonstrate that both designs suffer minor losses in their respective performances, validating the optimization procedure.
However, both designs feature bubble-like cavitation at the tip, which is problematic in terms of vibration and comfort. As it is suggested in *figure 11*, the cavity is attached to the nozzle shape, which indicates that the interaction between the propeller blade and the duct influences strongly the cavity size. Thus, the nozzle shape should be also optimized, which is not the case in this paper.

		Baseline	Optimum
Design condition 10km/h	*Kt*	0.58	0.58
	Kq	0.072	0.0705
	η	0.39	0.401
Off-design condition 11 km/h	*Kt*	0.506	0.505
	Kq	0.068	0.0667
	η	0.419	0.428
Weighted η		0.399	0.408
Cavitating volume [ml]		398	769

Table 7: Comparison of the baseline and the optimum performances.

6.2 NON-DETERMINISTIC OPTIMIZATION

6.2 (a) UQ simulations

An initial *UQ* study is performed with 13 uncertainties:

- 4 uncertainties for 4 different sections chord defined with a symmetric beta *PDF*;
- 4 uncertainties for 4 different profiles thickness defined with a beta non-symmetric beta *PDF*;
- 1 uncertainty for the rake, which represents the linear position of the tip compared to the root of the blade, defined with a symmetric beta *PDF*;
- 1 uncertainty for the gap between the blade and the duct, defined with a symmetric beta *PDF*;
- 1 uncertainty for the axial velocity defined in section *5.2(b)*.

By using a sparse grid level 1, 32 deterministic computations need to be run. Indeed, [15] proved a level 1 already provides accurate results on the mean and the variance, which is sufficient for the *RDO*.

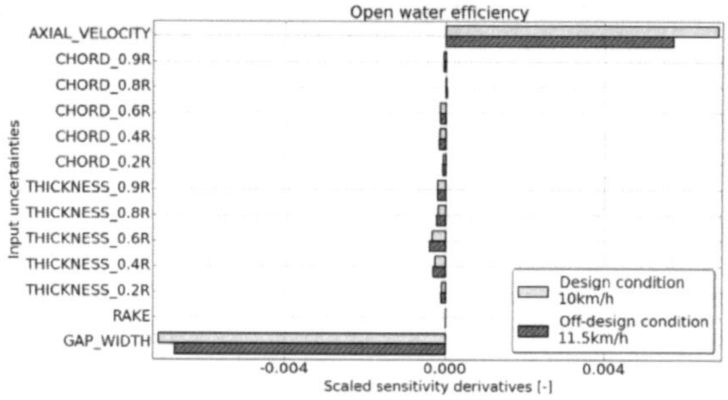

Figure 24: Scaled sensitivity derivatives of the open water efficiency for both operating points, with 13 uncertainties.

Scaled sensitivity derivatives (section *4.4(c)*) are then computed and show that the main influencing components are the axial velocity and the gap between the duct and the tip. This analysis is quite logical, because the efficiency η is function of the advance ratio of the propeller *J*, also function of the blade diameter and the axial velocity (cf *eq.5*).
Moreover, the uncertainties for the thickness having the lowest standard deviation, are clearly more important than those for the chord. This is also logical considering the fact the thrust and torque are directly influenced by each blade profile thickness.
From that analysis, it is decided to merge all chord and thickness uncertainties together. The change in the shape of the blade profile is controlled by a single section. And the rake uncertainty is neglected. This reduces the number of uncertainty parameters to 4, and the number of deterministic computations is now 10.

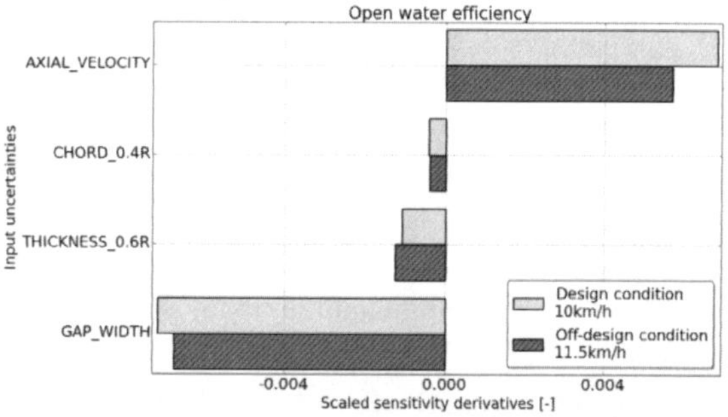

Figure 25: Scaled sensitivity derivatives of the open water efficiency for both operating points with 4 uncertainties.

The scaled sensitivity derivatives show a similar trend between 13 and 4 uncertainties (cf. *fig. 24 & 25*). Moreover, a normal distribution is reconstructed in order to display the difference in the mean and the variance (cf. *fig.26*).
It can be concluded that the reduction of the number of uncertainties did not impact the statistical moments results. Hence, 4 uncertainties are used during the *RDO*.

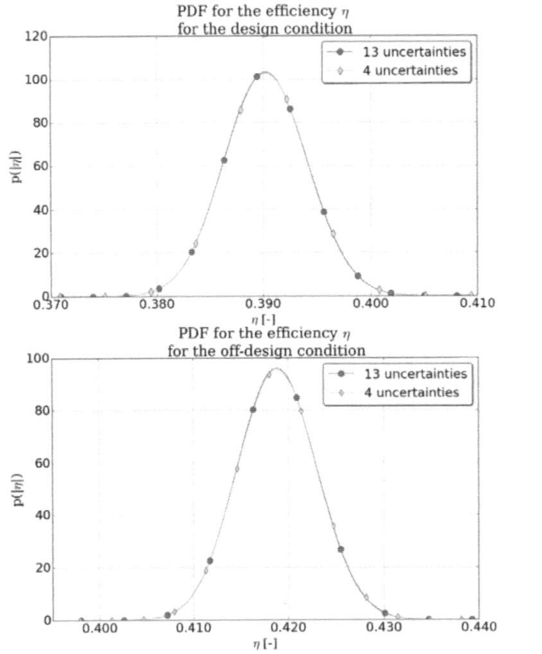

Figure 26: Normal distribution using the mean and standard deviation for both operating conditions and 13 and 4 uncertainties.

The scaled sensitivity analysis can also be useful for analysing the uncertainties influence over other key quantities, such as the cavitation volume, helping to identify critical variabilities. Indeed on *fig.27*, it can be seen that all geometrical and operating conditions strongly impact the cavity size, and might increase its volume by 4% when decreasing the axial velocity and 3.5% when increasing the propeller duct gap distance.

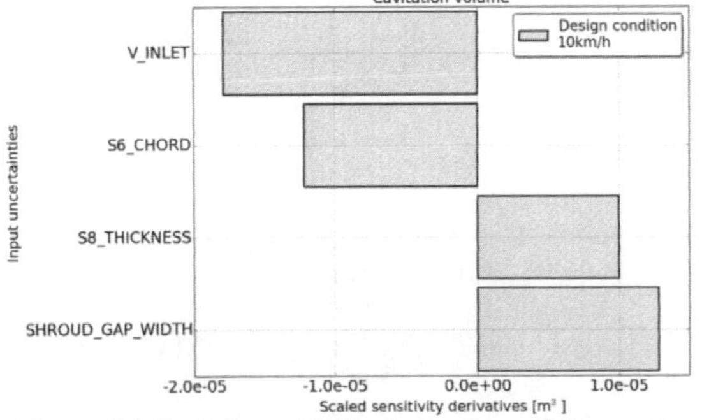

Figure 27: Scaled sensitivity derivatives of the cavity volume for the design condition.

6.2 (b) Robust optimization outcome and comparison with the deterministic optimum

The same design parameters are used, with 4 additional uncertainties for the *DoE* generation (cf. *section 6.2(a)*). The same operating points are considered. However, the formulation of the *RDO* is as followed:

- Improving the overall open water efficiency η by maximizing its mean value $\mu(\eta)$, the same aggregated formulation for both operating conditions;
- And also improving the robustness of the design by minimizing the standard deviation $\sigma(\eta)$;
- Maintaining the axial thrust, $\mu(K_t)$ is constrained and equal to *0.58*;
- Not deteriorating the performance, by keeping $\mu(\max volume)$ below *800 ml*.

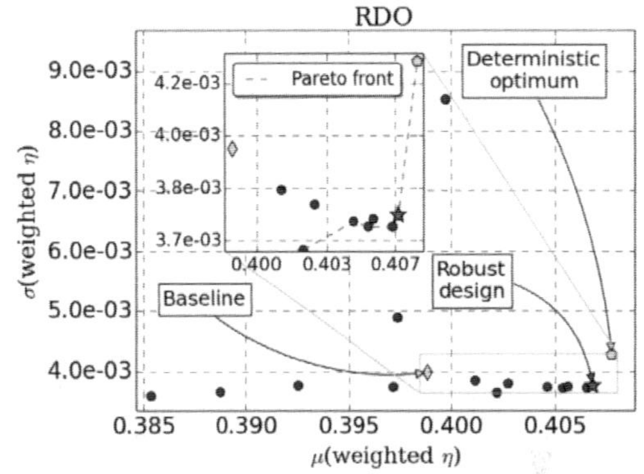

Figure 28: *RDO* history and Pareto front outcome.

In consequence, it can be seen that the *RDO* formulation is multi-objectives and the outcome is the Pareto front. The same number of iterations is used. For each iteration, 10 deterministic *CFD* computations are performed in order to extract the statistical moments.

Figure 28 clearly displays a robust design among the non-dominated samples, which manages to reduce the standard deviation by 6 % and the mean efficiency is increased by 2%. It is decided to select a single robust design, because it features the best trade-off in terms of statistical moments enhancement.

Moreover, a *UQ* study is performed on the deterministic optimum and shows an increase of the mean efficiency, but also an increase for the standard deviation of about 7.4% (cf. *fig.29*). Hence, the deterministic optimum is consequently less robust, but is still located on the Pareto front.

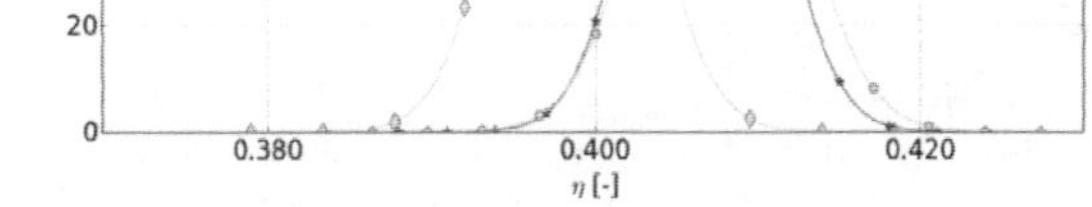

Figure 29: Normal distribution of the baseline, the robust and deterministic optima.

Moreover, the scaled sensitivity derivatives (cf. *fig.30*) reveal that the influence of the axial velocity and the diameter are still the largest.
However, the geometrical uncertainties show a discrete reduction in their respective influences for the robust design. This explains the decrease of the standard deviation.

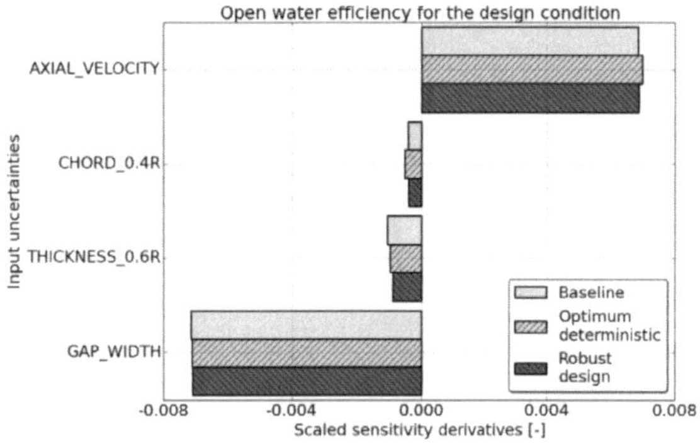

Figure 30: Scaled sensitivity derivatives at the design condition for the baseline, and the robust and deterministic optima.

Figure 30 shows a comparison of the propeller blade shapes for both designs, the deterministic and the robust. The robust design features a higher skewness in contrast to the deterministic optimum and the baseline (cf. *fig.16*). The explanation of the geometrical differences comes from the robust optimization formulation.
Indeed, the deterministic optimization aims at localizing the best design for a single objective formulation, whereas the *RDO* proposes a set of non-dominated designs with the objective to improve both performances and robustness.

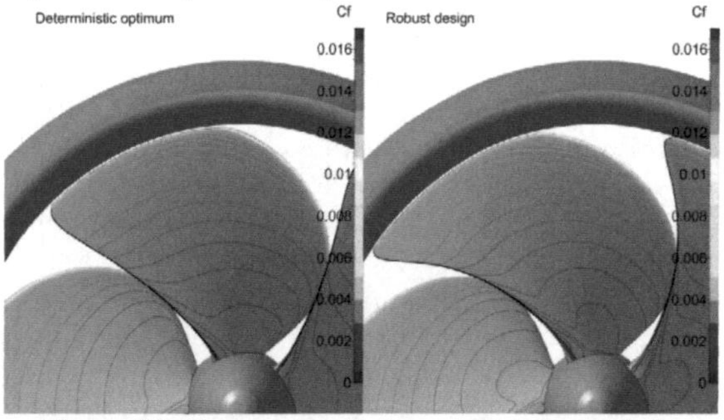

Figure 31: Comparison between the deterministic optimum (left) and the robust design (right).

7. CONCLUSION

In this paper, two design optimization studies were performed: a deterministic optimization and a robust optimization accounting for uncertainties. Since the investigated ducted propeller was subject to several operating conditions, two were considered: the design condition, which was used to draw the initial baseline geometry, and an arbitrary off-design condition based on the maximal cruising speed of the vessel.

As the cavitation was also of importance, a low *CPU* cost modelling based on the pressure field was used and provided interesting information.
Indeed, the first deterministic optimization indicated the relevance of considering cavitation modelling. Also, the outcome demonstrated the trade-off between performances improvements and cavitation criteria. Strictly speaking, the optimizer was restrained in the search of a better design due to the constraint in the cavitation volume. However, the performances can be further improved by including the nozzle shape parameters in future optimizations.
The second optimization was the robust design optimization that aimed at improving the performances while reducing the vulnerability of the design performances with respect to both operational and geometrical variabilities. The evaluation of those variabilities was performed by uncertainty quantification simulations. It was also observed that the *UQ* simulations can also play a major role in the design process for identifying critical manufacturing variabilities over key quantities of interest, such as cavitation or efficiency.
Eventually, both deterministic and robust optimum were compared, and as expected the robust design increased the performances and the robustness.
Moreover, the deterministic optimum was also located on the Pareto front, which might indicate that the inputted manufacturing variabilities were small enough to guarantee an acceptable stable behaviour, which was the expected results when using the most precise and costly S-class tolerance. Knowing the real-life performances can be assessed, the propeller designers can now try to apply coarser and less expensive manufacturing tolerance and use the proposed approach to optimize the design accounting for several variabilities and operating condition.

8. REFERENCES

1. INSIGHTS, GLOBAL MARKET, 'Inland Waterways Vessels Market to Hit $2.25 Billion by 2024.' *Global Trade Magazine, The Authority For US Companies Doing Business Globally.*, 30 Apr. 2018.

2. SIHN, WILFRIED, ET AL, 'A Green and Economic Future of Inland Waterway Shipping.', *Procedia CIRP vol 19 pp. 317–322*, 2015.

3. BUNDESANSTALT FÜR WASSERBAU, 'Driving Dynamics of Inland Vessels', *BAW*, 2016.

4. COLEY, DAVID A., 'An Introduction to Genetic Algorithms for Scientists and Engineers', *World Scientific*, 1999.

5. LOEVEN, G.J.A., WITTEVEEN, J.A.S., AND BIJL., H., "Probabilistic Collocation: An Efficient Non-Intrusive Approach for Arbitrarily Distributed Parametric Uncertainties", *45th AIAA*

Aerospace Sciences Meeting and Exhibit, Aerospace Sciences Meetings, 2007.

6. WUNSCH, D., NIGRO, R., COUSSEMENT, G., AND HIRSCH, C., 'Quantification of Combined Operational and Geometrical Uncertainties in Turbo-Machinery Design', *Proceedings of the ASME GT 2015*, 2015.

7. GOLUB, GH., WELSCH, JH., 'Calculation of Gaussian Quadrature rules', *Math. Comp. 23:221-230*, 1969.

8. SMOLYAK S., 'Quadrature and Interpolation Formulas for Tensor Products of Certain Classes of Functions', *Doklady Akademii Nauk SSSR Volume 4, pages 240-243*, 1963.

9. TURGEON ET AL., 'Parametric uncertainty analysis for thermal fluid calculations', *Nonlinear analysis 47 4533–4543*, 2001.

10. NIGRO, R., WUNSCH, D., COUSSEMENT, G., AND HIRSCH, C., 'Robust design in turbomachinery application', *Findings and Best Practice Collected During UMRIDA, a Collaborative Research Project (2013–2016) Funded by the European Union pp. 495-511*, 2018.

11. SALVATORE ET AL., 'Propeller Cavitation Modelling by CFD - Results from the VIRTUE 2008 Rome Workshop', *SMP'09*, 2009.

12. MERKLE ET AL., 'Application of Time-Iterative Schemes to Incompressible Flow', *AIAA Journal, Vol. 23, No 10, pp. 1518-1524*, 1985.

13. CARLTON, 'Marine Propellers and Propulsion', *Butterworth-Heinemann*, 2012.

14. ISO 484-2, 'Shipbuilding -- Ship screw propellers -- Manufacturing tolerances -- Part 2: Propellers of diameter between 0,80 and 2,50 m inclusive', 2015.

15. NIGRO, R., WUNSCH, D., COUSSEMENT, G., AND & HIRSCH, C., 'Uncertainty Quantification in Internal Flows', *55th AIAA Aerospace Sciences Meeting*, 2017.

9. AUTHORS BIOGRAPHY

Kevin Vidal holds the current position of marine *CFD* and optimization engineer at NUMECA International.

Leo Poppelier holds the current position of technical director at SIP Marine BV.

Benoit Mallol holds the current position of product manager and head of the marine application group at NUMECA International.

Charles Hirsch holds the current position of president and founder at NUMECA International.

AN ELECTRO-MECHANICAL CONTROLLABLE-PITCH PROPELLER

J McBain, Dogged Mechatronics LLC, USA
T Blakeley and **P R Eltringham,** Royal Institution of Naval Architects, UK

SUMMARY

An electro-mechanical replacement for hydraulically-actuated controllable pitch propellers is described. The proposed mechanism applies a thread to the rotating shaft driven by the ship's prime mover onto which a ferromagnetic nut is threadedly mated and spun by a stator-born electromagnet; the axial motion of the nut is used to manipulate the pitch of the rotatable blades. The design completely removes any mechanical coupling between the rotor and stator systems. An argument for the mechanism's efficiency, ease of maintenance, and reduction of weight is provided in the form of reasoning by analogy for the already well studied differentiation between hydraulic and electric cylinders. Efficiency gains of 2-3 times, a doubling of mean-time-between failure, and a reduction of weight by 10% are inferred.

1. INTRODUCTION

Controllable-pitch propellers (CPP) are an essential component to the drive train of marine and aerospace applications involving high manoeuvrability; rapid variations in propeller shaft speed call for corresponding changes in propeller pitch in order to maintain optimal thrust and maximal utilization of prime mover output. In the marine sector, fuel optimizations of 5-10% are commonly claimed [1-3] while some as high as 14-20% have been reported [4]. The means of actuating CPP is overwhelmingly based on hydraulic systems; essentially a hydraulic piston is embedded in the rotating shaft and hydraulic fluid is forced from off the rotating shaft into the rotating shaft causing the piston's motion which in turn rotates the propellers. Hydraulic cylinders have been the mainstay in linear actuation for most of the 20th century; however, electric cylinders have overtaken their hydraulic counterparts in many applications over the past decades. This trend begs the question why they have not been deployed in CPP applications.

The author proposes a replacement to hydraulic actuation of CPP with an electromagnetic apparatus found in electric cylinders used for linear actuation. The invention, advanced in US Patent 20180171973A1, purports to deploy a ferromagnetic nut threadedly mated to a thread on the propeller's rotating shaft; the nut is rotated by electromagnets on the stator system and its resultant axial motion is employed to manipulate the blades' pitch. The nut and electromagnets effectively amount to a reluctance motor employed for pitch manipulation – above and beyond whatever prime mover is employed to rotate the propeller's shaft. It will offer substantial efficiency over its hydraulic counterpart.

2. ELECTRO-MECHANICAL CONTROLLABLE PITCH PROPELLER

The entirety of the drive train is shown in Figure 1 with further detail of the interaction of the apparatus' components in Figure 2. A thread is applied to the shaft of the prime mover; a "ferro-magnetic nut" is threadedly mated to the prime mover's shaft and is spun by an electro-magnet – the pair of which creates a reluctance motor whose rotor system is free to move axially as constrained by the shaft's thread.

The electro-magnetic nut is rotatably mated to a pitch modulator which has a geared "rack" for each rotatable blade. The rotatable interaction of these two components is shown in Figure 3. The axial motion of the ferromagnetic nut pushes the pitch modulator back and forth in the blade hub which causes each geared "rack" to interact with the pinion of the rotatable blades leading the blades to rotate and change their pitch. This rack-and-pinion combination results in a highly efficient transfer of mechanical energy.

Control of the pitch modulation is almost the same as the control of a reluctance motor. A sensor will be required to measure the angular displacement of the ferromagnetic nut in order to switch current to each stator phase at the appropriate time. Once the reluctance motor's torque can overcome the mechanical friction of the thread, it can cause the ferromagnetic nut to rotate and actuate the blade's pitch. Closed loop control might be achieved by an additional linear displacement sensor measuring the ferromagnetic nut's axial displacement which would have a one-to-one correspondence to the pitch of the blades.

Notwithstanding the mechanical advantage already offered by the thread's chosen pitch, some CPP applications may require the application of torque much larger than a reluctance motor might generate. For those applications, the rotation of the "ferromagnetic nut" could instead be achieved by a gear and pinion pairing as shown in Figure 4. The axial modulating nut would have a spur gear profile on its exterior cylindrical surface and its rotation would be driven by a spur-gear in turn driven by some other rotational force. This arrangement offers the benefit of more choices of mechanical advantage but incurs additional weight and its control is less efficient.

Control of the augmented mechanism is somewhat unorthodoxed; steady-state is achieved by controlling the gear such that the tangential velocity at its gear teeth is equal to the tangential velocity of the axial modulator's gear teeth given by the angular rotation of the prime mover shaft and the axial-modulating nut's diameter. In this scenario, the axial-modulating nut remains stationary. In order to cause its axial motion in one direction or another, the relative speed under which the gear is driven would need to be increased or decreased. This would introduce an imbalanced torque on the axial-modulating nut causing it to rotate. Feedback control could be achieved by measuring the linear displacement of the axial-modulating nut.

3. EFFICIENCY

The Canadian Coast Guard recently conducted a study comparing the efficiencies of their electric and hydraulic winches. Their electric drive winch efficiency was found to be in the range of 70 to 85% while electro-hydraulic efficiency was found to be 45 to 70% and hydraulic drive was found to be 35 to 60%. [5] This provides a more objective basis for the commercial boasts of manufacturers like Tolomatic who claim that electric cylinders are twice as efficient as hydraulic cylinders. [6]

Much of the comparison between the two CPP approaches can be reduced to an analysis between hydraulic and electric cylinders – with one critical difference for this application: the seal between the rotor and stator systems needed for hydraulic CPP. Its oil distribution box, used to create this seal, is a source of substantial losses. A test bench was developed in [7] where it was found that "internal leakage is considered to influence the dynamics of the cylinder enormously" with measured external leakage of 0.5 ltr/min at 40 bar and internal leakage of 3 ltr/min at 100 bar. Given the parameters of this study, these leakages would seem to reduce the hydraulic CPP's efficiency to 25%. In marine applications, efficiency is further aggravated by power losses due to pressure drops in long shafts.

The potential to lock the propeller in place with a hydraulic valve would be appealing but for these power losses between the valves in the stator system and the cylinder in the rotor system. In contrast, the friction of acme-screw-type electric cylinders could provide a more efficient means of locking the pitch adjusting mechanism. Ordinarily such screws suffer from frictional efficiency losses between 30 to 50%; for applications not well suited to infrequent pitch manipulation such screws would not be much more efficient than hydraulics. However, for these scenarios demanding continual changes in pitch due to heavy manoeuvring, the 95% efficiency of ball or roller screws is required. [8] This approach would require additional energy to maintain the actuator at a constant position for those occasions demanding it. In the context of the proposed actuator, this would mean electric current would need to be provided to the magnets controlling the position of the nut/screw as opposed to the self-locking nature of the friction in the acme screw and the multiplicative mechanical advantage posed by the thread's pitch (requiring no further energy to keep the actuator stationary).

4. MAINTENANCE

During a period from 1969 to 1970, Drenth [8] gathered reliability statistics on controllable pitch propellers in marine applications. Studying 657 installations in service, all of which running 5,270,000 operating hours, 32 chance failures were noted – producing a mean-time-between failure of 165,000 hours and a reliability service period of 4000 hours. For a 95 % confidence level, the upper and lower limits of the calculated MTBF of 165,000 hours are 270,000 hours and 120,000 hours. Drenth further analysed the breakdown of failures and found that 26% of failures were related to hydraulics.

In contrast, Lemor [9] studied the reliability of roller screws in 14 supersonic airliners between 1976 and 1992 with 8 planetary roller screws operating the air intake ramps of the Rolls-Royce engines. In 1992, 62 "Air France" roller screws had performed an average of 7511 hours of operation without a single failure and 70 "British Airways" screws had performed an average of 10,377 hours. By 1996, all original roller screws remained in service resulting in a life performance of 11,690 hours. The effects were that their originally predicted estimate of 45,000 hours was a conservative one and that MTBF could be as high as 600,000 hours. The only periodic maintenance performed was cleaning of the exposed portions of the roller screw shaft and application of wide temperature grease; the periodicity of which was 200 to 230 hours. Reliability levels of 95%, 99%, 99.9% and even 99.95% are achievable with roller screws.

So far as the choice between roller screws and ball screws is concerned, Lemor deduced that since the predicted fatigue life of such a component is proportional to the cube of the dynamic rating, roller screws can outlive ball screws of comparable size by a factor of 25 to 30. Lemor also pointed out that the load rating of roller screws was three to four times that of a ball screw for screws of similar diameters.

The hydraulic CPP's reliability figures noted above are for the entire hydraulic system while the roller screw statistics do not include the entire actuating system in the proposed actuator. One must further consider the reliability of the reluctance motor driving the screw mechanism. Induction motors of this form are noted for their high reliability – in some cases exceeding 500,000 hours for a 50-kW motor. [9] While the motor driving the propeller shaft will have bearings, the reluctance motor driving the pitch control mechanism will not need bearings. The pitch-manipulating reluctance motor is just an electro magnet concentrically mounted to the ferromagnetic nut threadedly mated to the propeller shaft's thread; while electrical failures are still possible, as with any motor, the primary source of failures in motors tends to be their bearings which typically constitute half of the failures in such systems. [11]

The absence of bearings in the pitch-manipulating motor leaves the roller screw as the primary source of failure in the proposed actuating mechanism. Given the statistics already discussed, to say that the proposed pitch-manipulating actuator is 3 times as reliable seems conservative. This is consistent with the Canadian Coast Guard's finding that hydraulic systems have a substantial maintenance requirement over comparable electric systems. [10]

In [12], Qiao noted that actuator engineers estimated that electrical actuator systems could "achieve a significant reduction in the fuel burn and maintenance costs on an all-electric passenger plane and lead to a 30–50% reduction of the ground service equipment. For military aircraft in combat situations, the take-off weight could be cut down by 600–1000 lbs, and the vulnerability of the fuselage area could be reduced up to 14%." [12]

5. WEIGHT

While it is true that hydraulics "owns the trump card of power density" [13] where power transmission is concerned, such boasts are largely only true at the work point of the cylinder. One must further consider, the weight requirements of the heat exchanger and oil tank heater needed to allow the hydraulic CPP to operate in a variety of environments. Additionally, control valves, oil repositories, suction screens/filters, accumulators, gauges, hoses, fittings, etc. all add unnecessarily to the weight of hydraulic solutions. Experience while transitioning aircraft from hydraulic to all electric suggests that weight savings of 10% could be achieved from such upgrades. [14] Similar experiences were observed in offshore oil drilling redesign. [15]

The proposed CPP actuator constitutes a significant reduction in complexity. Its reduction in weight, as compared to hydraulics, should be at least as significant as that experienced in similar hydraulics-vs-electronics applications. Removing the hydraulic CPP's oil distribution box would bring additional weight savings not realized in past applications transitioning from hydraulic cylinders to electric cylinders.

7. CONCLUSIONS

A comparison of electric and hydraulic cylinders demonstrated that the proposed electromechanical CPP could be twice as efficient as that of its hydraulic counterpart; however, after considering the further inefficiencies introduced by the hydraulic CPP's oil distribution box, the factor of improvement could easily be triple. Similarly, based on present studies of the reliability of underlying components, the maintenance requirements of the electromechanical CPP is substantially less. The ugliness of preventive maintenance on hydraulic components could be replaced by infrequent addition of grease to the electromechanical CPP's screw; not much other attention would be needed and mean time between failure could be tripled. In aerospace contexts, a 10% weight saving could have an enormous impact on the energy efficiency and operational range which was not considered in the strict comparison between hydraulic and electric cylinders.

The simplicity of the proposed mechanism and its improvements in efficiency, reliability and weight savings define a new pareto optimal boundary for achieving controllable pitch propellers. It will allow the introduction of controllable pitch to simpler cost-sensitive applications like engine cooling fans, air conditioners, turbomachinery, etc.

9. REFERENCES

1. Kanome Propeller, 'A smart way of using a controllable pitch propeller', Retrieved from Kanome's website https://www.kamome-propeller.co.jp/en/brochures/pdf/kamome_smart_way_cpp_eng.pdf, 20 January 2019.

2. Rupp LA, 'CONTROLLABLE-PITCH PROPELLERS," Journal of the American Society for Naval Engineers, 61: 285-293. doi:10.1111/j.1559-3584.1949.tb05332.x

3. SKIPSREVYEn, 'Significant fuel savings by retrofit propeller solution,' Retrieved from SKIPSREVYEN website https://www.skipsrevyen.no/article/significant-fuel-savings-by-retrofit-propeller-solution/

4. Man Energy Solutions, 'Propeller and AftShip – Retrofit & Modernization,' Retrieved from MAN's website https://primeserv.man-es.com/marine-engines-and-systems/propeller-aftship/retrofit-modernization/propeller-upgrade

5. De Angelis V, 'Comparison Study of Electric, Electro-Hydraulic, and Hydraulic Drive Science Winches,' Retrieved from the European Research Vessel Operators website http://www.ervo-group.eu/np4/np4/%7B$clientServletPath%7D/?newsId=13&fileName=CCG_ERVO_Presentation_May_2009.pdf

6. Dietrich A, 'Electric rod actuators vs hydraulic cylinders: A comparison of the pros and cons of each technology," Retrieved from Tolomatic's website: https://www.tolomatic.com/Portals/0/Resources/89e5d4ac6565443aaa3bf8cdbedc45d1-orig.pdf

7. Dullens F: Modelling and Control of a Controllable Pitch Propeller, Technische Universiteit Eindhoven, February 2009. Available on the web: http://www.mate.tue.nl/mate/pdfs/10299.pdf

8. Drenth BW, 'Reliability of Controllable Pitch Propellers,' International Shipbuilding Progress, 1 Jan 1972, vol 19 no 220 pp. 415-425.

9. Lemor PC, 'The roller screw, an efficient and reliable mechanical component of electro-mechanical actuators,' IECEC 96. Proceedings of the 31st Intersociety Energy Conversion Engineering Conference, Washington, DC, USA, 1996, pp. 215-220 vol1 doi: 10.1109/IECEC.1996.552873

10. Leroy Somer, 'Three-phase induction motors for rail applications 01 to 50 kW,' retrieved from Leroy Somer's website: http://acim.nidec.com/motors/leroy-somer/products/induction-motors/definite-purpose/fd-fe-fs

11. Ferriera FJTAE et al, 'Reliability and Operation of High-Efficiency Induction Motors,' in EEE Transactions on Industry Applications, vol 52 no 6 pp 4628-4637, Nov-Dec 2016. Doi: 10.1109/TIA.2016.2600677

12. Qiao et al, 'A review of electromechanical actuators for More/All Electric aircraft systems,' Proceedings of the Institution of Mechanical Engineers, Part C: Journal of Mechanical Engineering Science, 232(22), 4128-4151, https://doi.org/10.1177/0954406217749869

13. Cosford J, 'Is it a fair fight? Hydraulics vs electrics,' Retrieved from Mobile Hydraulic Tips' website https://www.mobilehydraulictips.com/fair-fight-hydraulics-vs-electrics/

14. Helsley C. 'Power by Wire for Aircraft – The All-Electric Airplane,' SAE Technical Paper 771006, 1977, https://doi.org/10.4271/771006

15. Pawlus et al, 'Hydraulic vs Electric: A Review of Actuation Systems in Offshore Drilling Equipment.' Modelling, Identification and Control: A Norwegian Research Bulletin 37. 1-17, 10.4173/mic.2016.1.1.

9. AUTHORS BIOGRAPHY

Jordan McBain is an independent contracting engineer in Detroit's automotive sector. He provides services in the field of mechatronics with a focus on software engineering based on his core academics in a Bachelor of Engineering and Management attained at McMaster University in Software Engineering in 2005. Having served several years as a Naval Combat Systems Engineering Officer in the Royal Canadian Navy, with operational deployments in HMCS REGINA and shore postings serving the Canadian Fleet, he released from the Canadian Forces to complete a Ph.D. in Natural Resources Engineering developing fault detection techniques for transmissions operated under varying speed and load at Laurentian University in July 2012. He has worked in the US automotive sector in various roles involving controls, model-based design and software engineering.

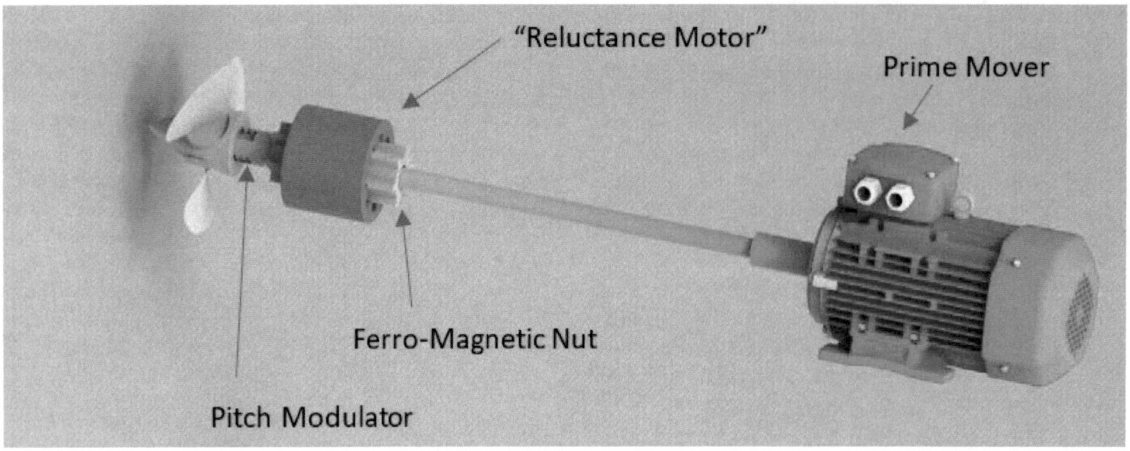

Figure 1 Electro-mechanical CPP drivetrain

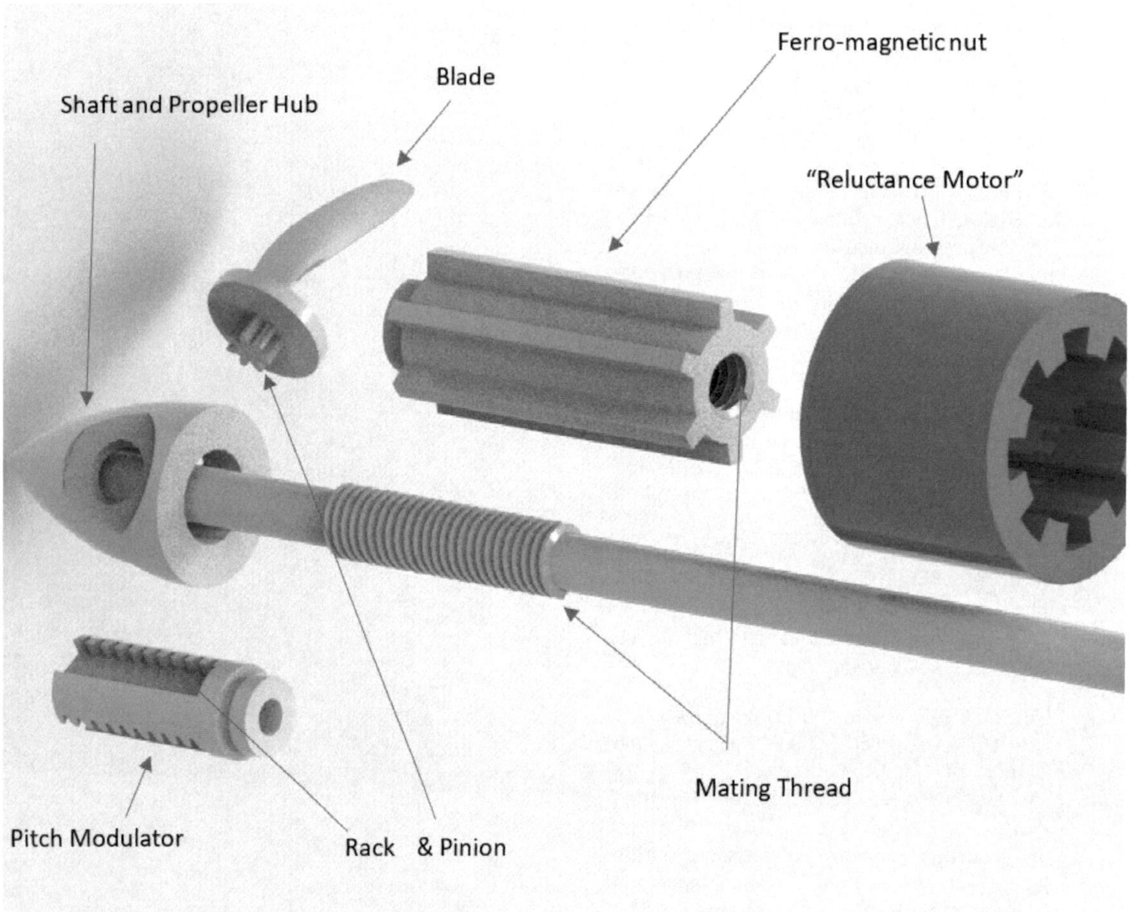

Figure 2 Electro-mechanical CPP components

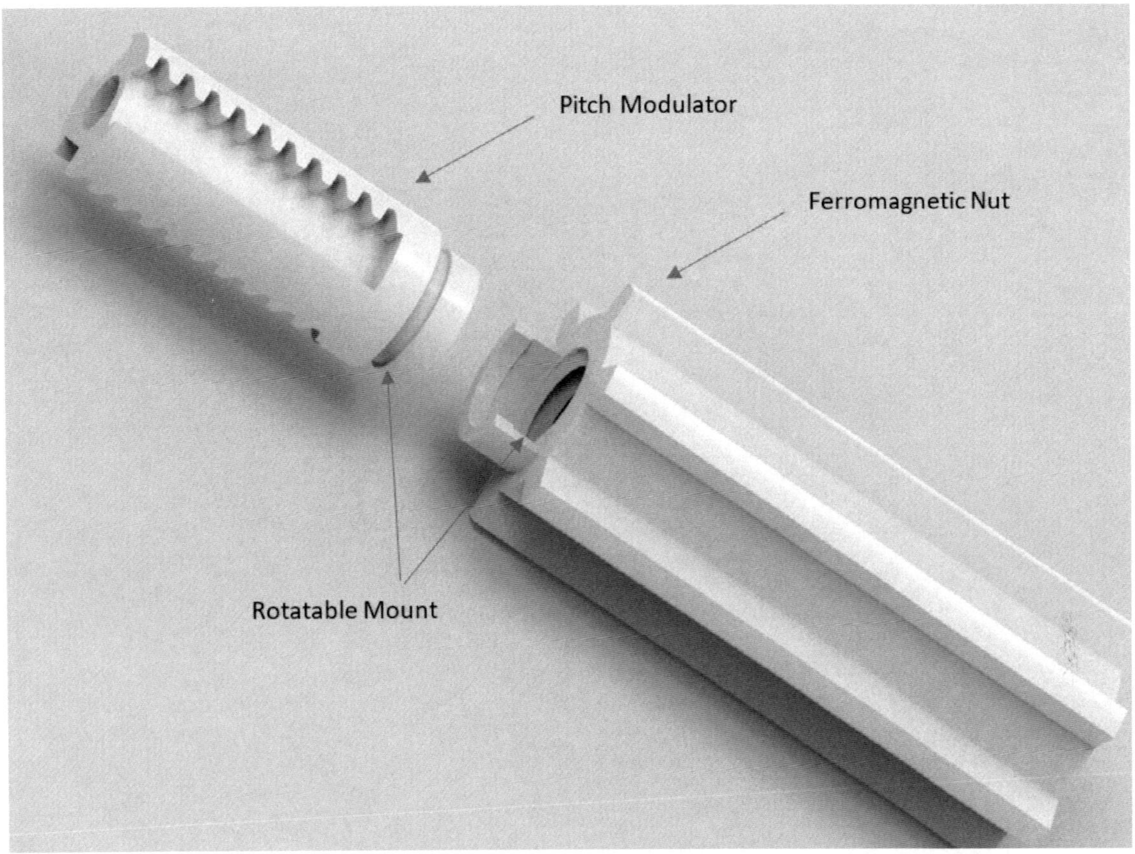

Figure 3 Rotatable interaction of pitch modulator and ferromagnetic nut

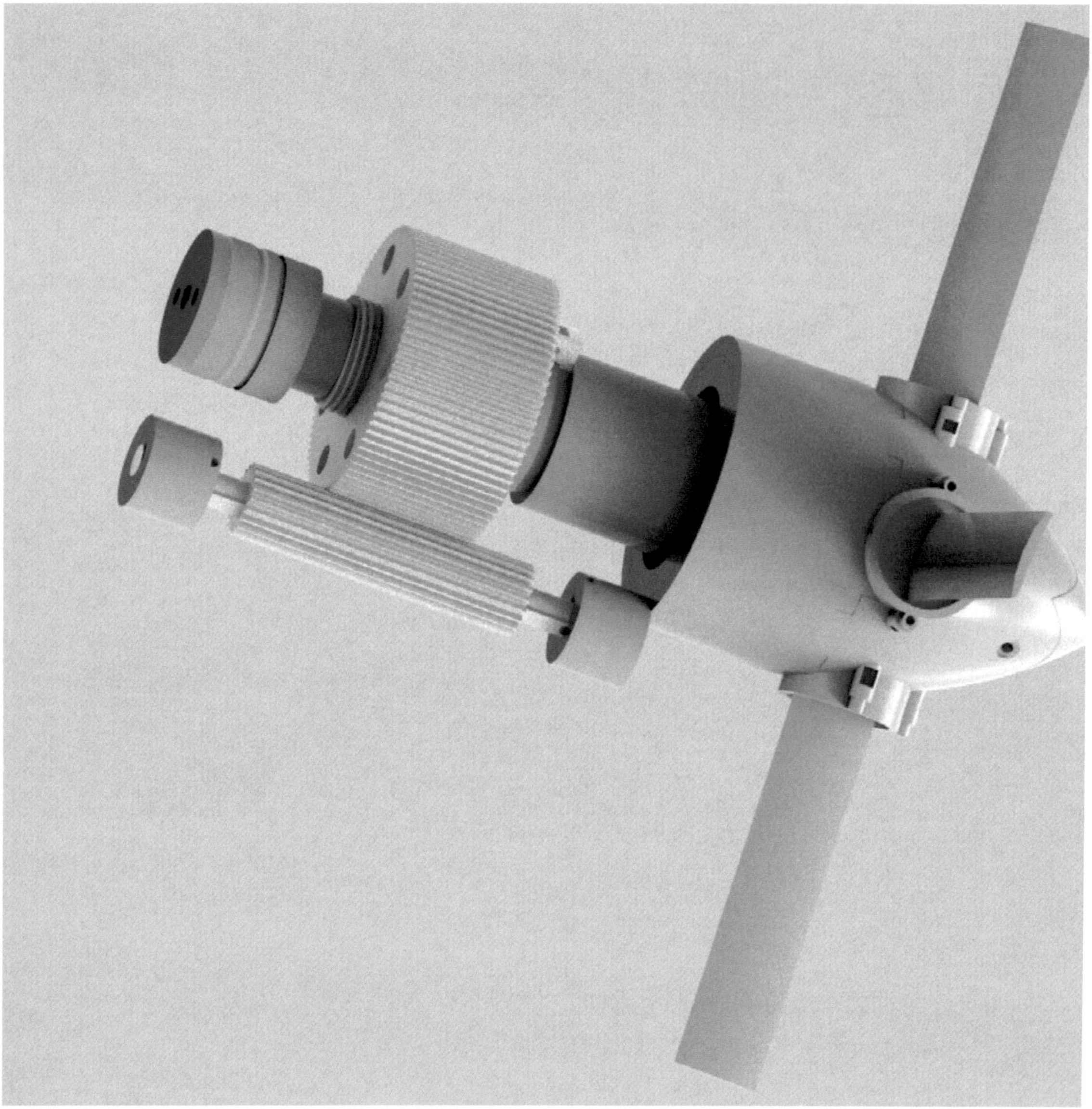

Figure 4 Gear-driven axial modulator

FULL SCALE PERFORMANCE OF GATE RUDDER

N. Sasaki, University of Strathclyde, UK
S. Kuribayshi, Kuribayashi Steamship Co., Japan
A. Miles, Stone Marin Propulsion, UK

SUMMARY

This paper introduces full scale performance results obtained from voyage data taken from two 2400 dwt container ships, one equipped with the world's first GATE RUDDER®. This is an innovative design, which is not seen as a conventional energy saving device and which has not been fully explored so far. The recent full-scale trials with this coastal container vessel have confirmed the superior performance of the GATE RUDDER® system which has shown 14% reduction in fuel consumption when compared with it's sister vessel having a conventional rudder system. After 12 months from delivery, the voyage data from both vessels revealed that in some situations the gain could be much larger at abt. 27%. In this paper, the full scale operational performance differences from not only the model test but also the sea trial result are presented.

NOMENCLATURE

C_T	Propeller loading factor (-)
DWT	Deadweight (ton)
H_W	Significant Wave Height (m)
P_0	Power predicted by Tank Test (kW)
P_{ACT}	Power measured in service (kW)
R_0	Hull resistance at calm sea (N)
R_A	Additional resistance (N)
R_{AA}	Additional resistance due to wind (N)
R_{AW}	Additional resistance due to waves (N)
R_{AR}	Additional resistance due to rudder (N)
R_H	Increased hull resistance due to prop. (N)
R_{RC}	Resistance of conventional rudder (N)
R_{RG}	Resistance of gate rudder (N)
t	Thrust deduction factor (-)
T_P	Propeller thrust (N)
T_{PC}	Propeller thrust of Conv. Rudder case (N)
T_{PG}	Propeller thrust of Gate Rudder case (N)
V_W	Relative wind velocity (m/sec)
w_E	Effective wake factor (-)
w_N	Nominal wake factor (-)
w_{NF}	Frictional part of nominal wake factor (-)
w_{NP}	Potential part of nominal wake factor (-)
w_R	Potential rudder wake factor (-)
κ	Resistance ratio to hull resistance (-)
κ_P	Reduction ratio of main prop. thrust(-)
ψ	Relative wind direction (deg.)

1. INTRODUCTION

Rudders are primarily applied on conventional ships for course keeping and manoeuvring. The fundamental studies have been published to investigate the rudder performance mainly from the aspect of manoeuvring.[1)2)] There are a few studies investigating the optimum rudder design for propulsive performance which give us useful information about the rudder positions such as longitudinal position, lateral position and vertical position etc..[3)4)5)]

Photo 1. Conventional Flap Rudder (upper) vs. Gate Rudder(lower)

Table 1 Principal dimensions of Sakura (Flap Rudder) & Shigenobu (Gate Rudder)

	Sakura	Shigenobu
Loa (m)	111.4	
B (m)	17.8	
d (m)	5.24	
Main Engine	3309kW x 220rpm	
Rudder	Flap Rudder	Gate Rudder
Delivery	August 2016	December 2017

However, a ship's rudder is one of the sources in contributing to the ship's resistance when it is located behind the propeller. Within this context, the main purpose of the Gate Rudder® system is to remove this source or rather replace with a thrust source (like a duct) in order to reduce the required main propeller, thrust and hence reduction of required engine power. With this idea, the rudder may become an ESD by being placed aside the propeller, instead of behind the propeller, to simulate the duct effect of a ducted propeller.

The Gate Rudder® has two rudder blades with asymmetric sections, which are located adjacent to the propeller, and each blade can be controlled independently. The two rudder blades, encircling the propeller at the top and sides, provide a duct effect similar to ducted propeller and hence produce additional thrust, as opposed to the additional drag of a conventional rudder behind the propeller.[6)7)] Owing to this additional thrust by the Gate Rudder, the required thrust of the propeller as a main propulsor can be reduced more than 10%.

By introducing this ESD based on the simple idea, the interaction among propeller, hull and conventional rudder can be replaced by a completely different interaction scenario. These differences between the conventional rudder and the Gate Rudder systems are analysed theoretically in section 2 and 3.

Photo 1 and Table 1 show a comparison of the conventional rudder and the Gate Rudder® system on two sister vessels which are the subject of this paper. The independent control of the two rudder blades also provides effective control of the propeller slipstream and hence steering, thus the Gate Rudder® system presents not only more propulsive efficiency but also greater manoeuvrability. In addition to these two major advantages of the Gate Rudder® system, there are other performance superiorities, which are noticed based on further analysis of the voyage data, including reduced resistance during seakeeping performance as well as manoeuvring motion. Such preferred full scale features are explained in section 4 and 5.

2. GATE RUDDER; HOW IT WORKS

2.1 INTERACTION AMONG RUDDER, PROPELLER AND HULL IN CALM SEA

2.1(a) Conventional Rudder System

The propeller produces a much large thrust than the towed hull resistance. This is due to propeller action near the stern.

This scenario can be called vicious circle as shown in Figure 1 because the increase in resistance is proportional to propeller thrust and consequently the propeller thrust is making additional resistance due to this interaction.

The increased resistance ΔR due to propeller action and rudder is defined as follows;

$$\Delta R = \Delta R_H + \Delta R_{RC} \qquad \text{-------- (1)}$$

Where, ΔR_H is increased resistance due to interaction between the propeller and hull, and ΔR_{RC} is resistance of conventional rudder during propelling condition.

Assuming the hull resistance R_0 at towing condition as 100%, ΔR_H and ΔR_{RC} are ranged 15%-25% and 3%-7% respectively depending on stern geometry and rudder type and geometry. Hence, the total increase in resistance will be around 20%-30% in comparison with the hull resistance which was assumed 100%.

This interaction is called thrust deduction and the ratio to the total increase in resistance to the propeller thrust is called thrust deduction factor (t_H).

The thrust deduction factor (t_H) can be measured by the self-propulsion test and ranges from 0.15-0.23 when the resistance test is conducted without the rudder. The thrust deduction factor shows sometimes shows very low values such as 0.12 to 0.08. There are two reasons for this. One possibility is the case when the flow separation at the stern is significant when the propeller is not working. In this case, the resistance of original hull is extremely high due to this separation, however this does not happen during propelling conditions. Therefore, the thrust reduction factor does not indicate accurate values because it contains the effect of flow separation occurring from an unrealistic flow field during the resistance test. Another case is when the propeller is far from the stern boss end and the rudder is not behind the propeller plane. This can be seen in the case of twin screw vessels, and which is not our subject here.

For the typical example, the total increase in resistance due to the propeller action is 25% when t_H =0.20. This means that the required propeller thrust is almost 25% higher than the hull resistance at towing condition. If the rudder resistance ΔR_{RC} can be assumed 5% of the hull resistance, ΔR can be simply calculated at 30% of the hull resistance.

It is well known fact that thrust deduction factor (t_H) is constant with the variation of the propeller thrust. This means ΔR_H will increase (decrease) proportional to the increase (decrease) of the propeller thrust.

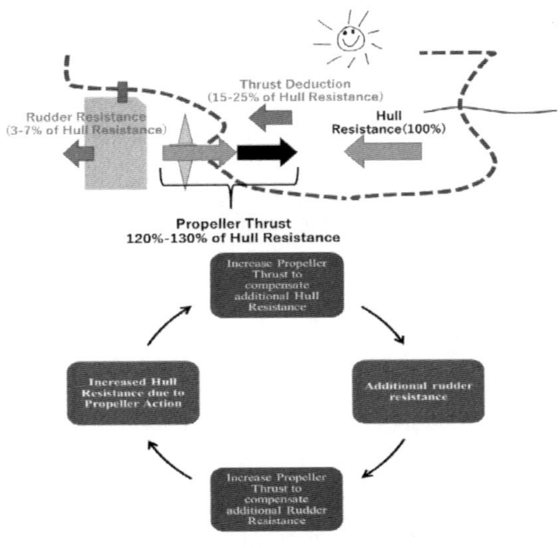

Figure 1 Vicious Circle of Conventional Rudder System (Calm Sea)

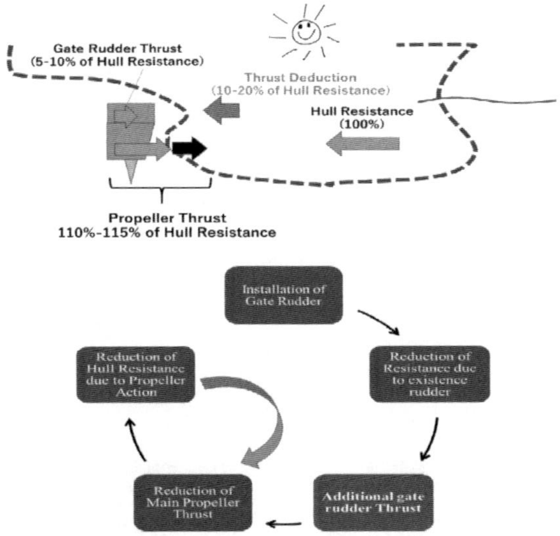

Figure 2 Virtuous Circle of Gate Rudder System (Calm Sea)

The propeller thrust for the conventional propeller system can be calculated from the following equation;

$$T_P = R_0/(1 + \kappa_C - t_H) \text{ -------- } (2)$$

Where, κ_C is rudder resistance ratio to ship hull and given by

$$\kappa_C = -\Delta R_{RC}/R_0 \qquad \text{-------- } (3)$$

2.1(b) Gate Rudder System

By removing the conventional rudder stern and placing the gate rudder adjacent to a propeller as a combined propulsor, we can expect an evolutionary change in the interaction among the rudder, propeller and ship hull.

The main difference is the required thrust of the screw propeller. The propeller thrust can be reduced considerably as shown in Figure 2.

It is noted that the interaction between the hull and Gate Rudder is negligibly small which was reported by Sasaki[7)].

It seems that interaction can occur within the stream tubes which are affected by the propulsor comprising not only a propeller but also rudder blades. The distance between the gate rudder blades is far from the hull within these stream tubes.

The increased resistance ΔR due to propeller action is defined like the conventional rudder case;

$$\Delta R = \Delta R_H + \Delta R_{RG} \qquad \text{-------- } (4)$$

However, there are many differences we should consider.

The first, ΔR_{GR} is negative because the Gate Rudder is producing thrust instead of resistance. The second, ΔR_H is much smaller than the case of equation (1).

We have two reasons for smaller ΔR_H.

One is the fact that the main propeller thrust is smaller than for the conventional rudder case. The main propeller thrust can be calculated by the following equation;

$$T_P = R_0/ (1 + \kappa_G - t_H) \qquad \text{------ } (5)$$

Where, κ_G is rudder thrust ratio to hull and given by

$$\kappa_G = -\Delta R_{RG}/R_0 \qquad \text{-------- } (6)$$

Comparing equation (2) to equation (5), the reduction ratio (κ_P) of the propeller thrust can be obtained easily as shown in Figure 3 for the typical case of $t_H = 0.18$.

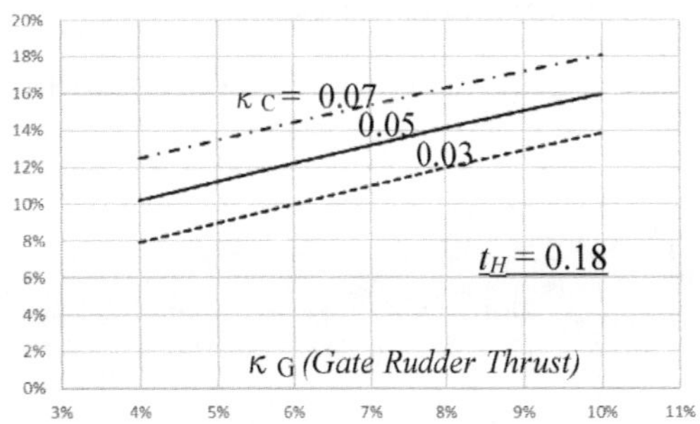

Figure 3. Thrust Reduction Ratio κ_P

Figure 3 shows the fact clearly that the Gate Rudder system can reduce the required main propeller thrust considerably.

The second reason for smaller ΔR_H is the fact that the interaction between the Gate Rudder blades and the hull is negligibly small.

This fact was found at Newcastle University when the force measurement for the floating stern model was conducted. Photo 2 shows the model ship with floating stern section. It was revealed that the increased resistance on the stern part is only dependent on the main propeller and the effect of the Gate Rudder thrust was negligibly small [7)].

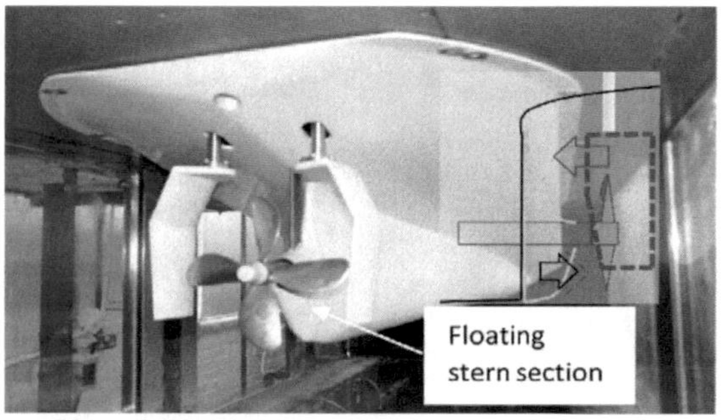

Photo 2 Floating Stern Model to investigate the Interaction among Propeller, Rudder and Stern

It is noted that the average values for κ_C and κ_G is 4% and 6% respectively for the model scale. This implies that the reduction ratio of the main propeller thrust (κ_P) by the Gate Rudder system at calm sea will be expected to be more than 10% according to the Figure 3.

Here, κ_P is the reduction ratio of the main propeller thrust for the conventional rudder case (T_{PC}) and the gate rudder case (T_{PG}) respectively, and represented by

$$\kappa_P = (T_{PC} - T_{PG}) / T_{PC} \text{ -------- } (6)$$

It seems that κ_P is the most important parameter for prediction power savings with the Gate Rudder.

2.2 INTERACTION AMONG RUDDER, PROPELLER AND HULL AT ROUGH SEA

The increased resistance ΔR due to propeller action at rough sea condition is defined as follows;

$$\Delta R = \Delta R_H + \Delta R_R + \Delta R_A \quad \text{-------- } (7)$$

Where, ΔR_A is the increased resistance due to wind, waves and rudder steering to keep the ship heading constant. Hence, ΔR_A can be represented by

$$\Delta R_A = \Delta R_{AW} + \Delta R_{AA} + \Delta R_{AR} \quad \text{-------- } (8)$$

Where, ΔR_{AW} and ΔR_{AA} are the added wave and added wind resistance respectively. ΔR_{AR} is the increased rudder resistance which originated from following reasons;

1) Increased propeller loading (higher slip stream)
2) Rudder helm to keep the course constant

The second contribution has an extremely large impact on the resistance compared with the first one. The increased rudder drag due to increased propeller loading is not so large because the drag can be compensated by higher leading edge suction so called.

On the contrary, the increased rudder resistance due to higher helm activity shows almost the same level to added resistance as from waves and wind, which are unavoidable.

This will bring us further detrimental effects in the "vicious circle" as shown in the Figure 4.

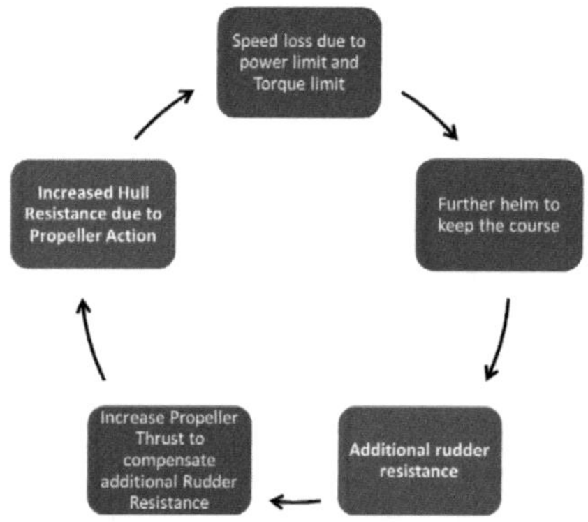

Figure 4 Vicious Circle of Conventional Rudder System due to Rough Weather

On the other hand, the Gate Rudder system shows the remarkable reduction in the rudder resistance during helm action.

Figure 5 shows this advantage and the model test revealed that the Gate Rudder system has a strong possibility to reduce the sea margin owing to such favourable characteristics. The measurement may not be so accurate because of the low Reynold number of the test conditions, however the difference is very clear.

This remarkable change in rudder drag was also confirmed from the manoeuvring test results during her sea trial. It appeared on her higher ship speed during the manoeuvring motion.

Photo 3 Test Set up for Rudder Force and Hull Force Measurements

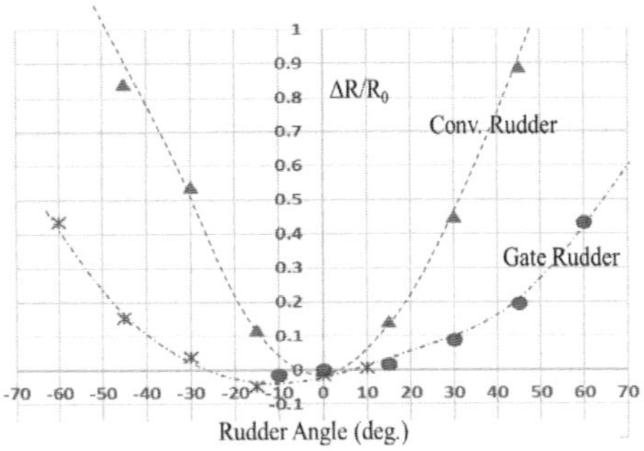

Figure 5 Hull Resistance Increment Ratio introduced by the Rudder Helm

Figure 5 shows the measured rudder resistance during the helmed condition with propeller rotation using the 2.5 m model shown in Photo 3

Taking the difference of this rudder resistance into account, the required main propeller thrust in rough sea condition can be represented as follows:

$T_{PC} = (R_0 + \Delta R_{AW} + \Delta R_{AA})/(1 + \kappa_{CA}(\delta) - t_H)$ *for Conv. Rudder* -------- (9)

$T_{PG} = (R_0 + \Delta R_{AW} + \Delta R_{AA})/(1 + \kappa_{GA}(\delta) - t_H)$ *for Gate Rudder* -------- (10)

Here, the difference of ΔR_{AW} and ΔR_{AA} between the rudder systems is neglected,

From the characteristics of wing, $\kappa_{CA}(\delta)$ and $\kappa_{GA}(\delta)$ can be represented by following formula respectively:

$$\kappa_{CA}(\delta) = C_1\ \delta^2 + \kappa_C \quad \text{-------- (11)}$$

$$\kappa_{GA}(\delta) = C_2\ \delta^2 + \kappa_G \quad \text{-------- (12)}$$

Where, C_1 and C_2 can be obtained from Figure 5 as 1.559 and 0.267 respectively, however these values will vary depending on the rudder configuration.

3. WAKE SCALING

The conventional full scale wake prediction is based on following two ideas;

(1) The inflow at the propeller plane will be increased by a thinner boundary layer than that of model scale due to higher Reynolds number.
(2) The propeller inflow will be increased further due to the effect of propeller action

The wake flow representing effect (1), only is called nominal wake and the wake including effect (2) is called the effective wake.

Therefore, the difference between nominal and the effective wake originates from the deformation of the boundary layer which is caused by propeller suction. It is obvious that this deformation strongly depends on the propeller suction which is almost proportional to propeller thrust.

This phenomenon can be seen frequently during the propeller loading variation in the propulsion test. Therefore, the full scale effective wake can be represented as follows;

$$w_{EM} = w_{NM} - \delta w(C_{TM}) \quad \text{-------- (13)}$$

$$w_{ES} = w_{NS} - \delta w(C_{TS}) \quad \text{-------- (14)}$$

Where $\delta w(C_T)$ is a linearized deformed wake function for a propeller loading factor given by

$$\delta w(C_T) = C^* * C_T \quad \text{-------- (15)}$$

C^* can be obtained from the result of a propeller loading variation test during the propulsion test.

Because w_N contains a potential wake originating from the ship stern and rudder, w_N can be replaced by

$$w_N = w_{NF} + w_H + w_R \quad \text{-------- (16)}$$

Neglecting the difference of relative boundary thickness between the model and ship, w_H and w_R can be regarded as the same value for the model scale and full scale.

Based on this idea, the ITTC wake scaling procedure is introduced as follows;

$$w_{NS} = ((1+K)C_{FS} + \Delta C_F)/(1+K)C_{FM} * w_{NF} + w_H + w_R \quad \text{-------- (17)}$$

$$w_{ES} = w_{NS} - \delta w(C_{TS}) \quad \text{-------- (18)}$$

Assuming

$$\delta w(C_{TS}) = ((1+K)C_{FS} + \Delta C_F)/C_{FM} * \delta w(C_{TM}) \quad \text{---- (19)}$$

w_{ES} *can be represented by*

$$w_{ES} = ((1+K)C_{FS}+\Delta C_F)/C_{FM} * w_{EM} + w_H + w_R \text{---- (20)}$$

Assuming the sheer stress part of the thrust deduction factor is negligibly small,

$$w_H = tp = t \quad \text{-------- (21)}$$

Finally,

$$w_{ES} = ((1+K)C_{FS}+\Delta C_F)/C_{FM} * w_{EM} + t + w_R \text{----- (22)}$$

Through the above procedure, there are two parts where we should take the difference into account as follows for the Gate Rudder case;

$$w_R = 0 \quad \text{-------- (23)}$$

$$\delta w(C_T) = C^* * C_{TG} \text{ -------- (24)}$$

Where,

$$C_{TG} < C_{TS} < C_{TM} \quad \text{-------- (25)}$$

4. PREDICTION AND SEA TRIAL RESULTS

4.1 PREDICTION OF FULL-SCALE GATE RUDDER PERFORMANCE

Delivered power (P) at ship speed (V_S) can be obtained from following equations;

$$P = T_P * V_S*(1-w_{ES}) / \eta_P \text{ -------- (26)}$$

η_P can be estimated based on propeller-rudder open water characteristics of the designed propeller with a gate rudder.

If the propeller-rudder open water test is not available, the following empirical formula can be used for the first stage estimation.

$$\eta_P = \alpha*2/(1+(1+C_T)^{0.5}) \text{ -------- (27)}$$

α is coefficient and around 0.78-0.82 for a conventional propellers.

For the case of the container ship shown in Table 1, the power curve will be obtained by the following equations using the model test data.

$$P = T_P * V_S*(1-w_{ES}) / \alpha*2/(1+(1+C_T)^{0.5}) \text{ ---- (28)}$$

$$T_P = R_0/(1 + \kappa_G - t_H) \text{ -------- (29)}$$

$$w_{ES} = ((1+k)C_{FS}+\Delta C_F)/(1+k)C_{FM} * w_{EM} + t_H \text{ --- (30)}$$

For fully loaded conditions;

$t_H = 0.113$
$w_{EM} = 0.369$

$w_{ES} = 0.340$

$\kappa_G = 0.023$ *(model scale)*

For sea trial conditions;

$t_H = 0.140$

$w_{EM} = 0.373$

$w_{ES} = 0.364$

$\kappa_G = 0.023$ *(estimated)*

Using these data obtained from the model test, estimated power curves are compared with full scale data in next section.

4.2 SPEED TRIAL DATA

The speed trial was conducted according to the normal procedure for a shipbuilder which is three sets of double runs. The ground speeds by DGPS were recorded and power and fuel consumptions were calculated by engine parameters such as LI and Pmax, and flow meter respectively.

The obtained power (Figure 6) was examined from several aspects, however the most important aspect was considered to be fuel consumption and it was concluded that the fuel oil consumption ratio of the engine (g/kw,hrs) of both vessel was almost the same.

Table 2 and Table 3 show the raw data for the speed trials for two vessels.

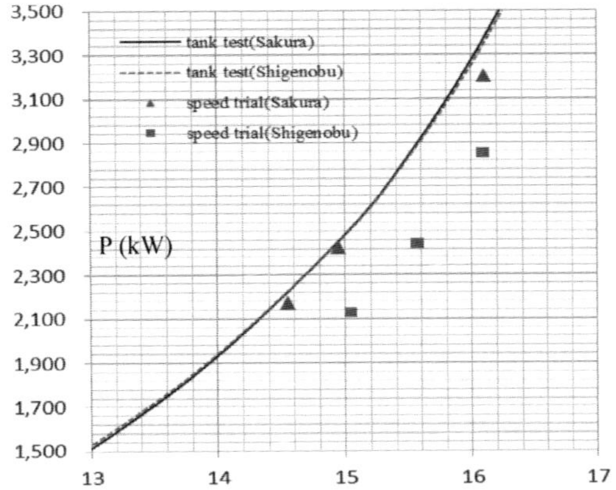

Figure 6 Tank Test and Speed Trial Results of two Ships

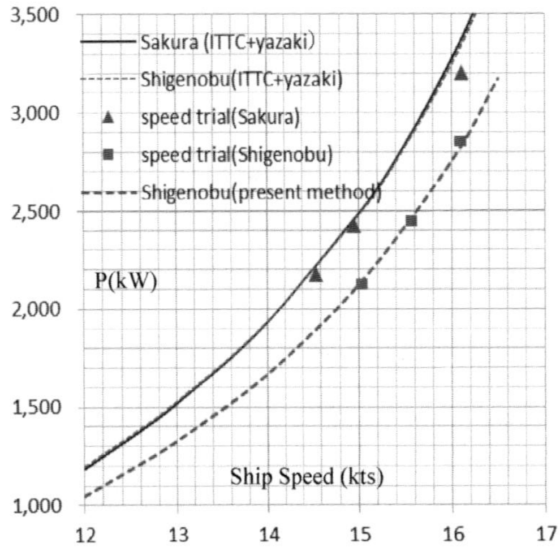

Figure 7 New Power Prediction Method

Table 2 Speed Trial Result (Conventional Rudder)

	Speed(kt)	Power(kw)	Vw(m/s)	φ(deg)	Hw(m)
1st run	13.433	2230	5.0	0.0	0.2
2nd run	15.636	2230	8.0	10.0	0.2
3rd run	14.107	2535	5.0	9.0	0.3
4th run	15.745	2535	10.0	10.0	0.3
5th run	15.455	3275	7.0	0.0	0.1
6th run	16.634	3275	10.0	10.0	0.1

Table 3 Speed Trial Result (Gate Rudder)

	Speed(kt)	Power(kw)	Vw(m/s)	φ(deg)	Hw(m)
1st run	16.174	2430	8.0	-20	0.75
2nd run	14.679	2490	11.0	20	0.75
3rd run	16.2	2705	6.0	-30	0.75
4th run	15.435	2749	13.5	10	0.75
5th run	16.426	3256	3.5	-10	0.75
6th run	16.416	3362	13.0	10	0.75

Figure 6 shows the analysed data of the two ships and the difference in power at the same speed is 14%. This is enormous because we have never seen such an evolutionary energy saving device since the event of the competition between HMS Alecto and HMS Rattler.(say which is prop & which a Paddle)

As explained in the sections 2 to 3, the powering procedure and wake scaling of the gate rudder case is different from a conventional rudder case. Using the conclusions derived from section 4.1, power obtained from the sea trial is compared to the prediction as shown in Figure 7.

In addition to the power, the effective wake was investigated. Because the difficulty in measuring the gate rudder thrust, the present analysis was made based on the K_Q identity which is derived from propeller torque. This value is affected by the flow acceleration due to the gate rudder blades and corresponds to conventional powering procedures.

Therefore, we have only possibility for the validation of the accuracy of conventional powering procedures for wake scaling. The result is very interesting. In the Figure 8, it is very obvious that the prediction of the effective wake based on the conventional way is not working properly.

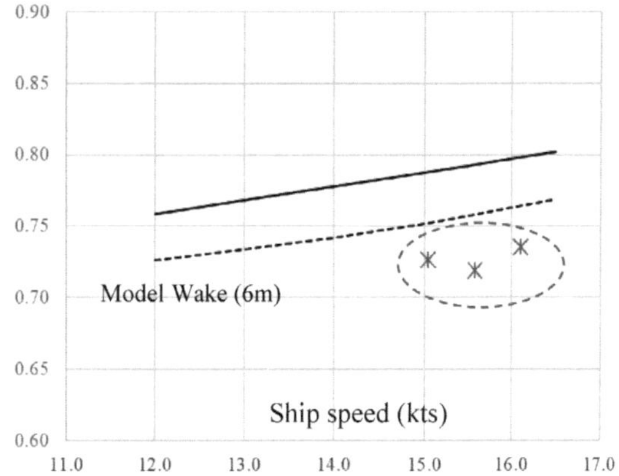

Figure 8 Conventional Wake Scaling Method and Sea Trial Result(GR)

The reason of this discrepancy is very clear. The existing wake scaling is not able to distinguish the actual effective wake and rudder induced velocities. If we want to use the conventional powering procedure, we need to separate the obtained propeller advance speed into these two components, i.e., actual effective wake and rudder induced wake. If the ship stern and the propeller diameter are similar for two rudder configurations, the gate rudder induced wake can be roughly estimated by the difference of two obtained propeller advance speeds.

4.3 MONITORING DATA AFTER DELIVERLY

As explained in the section 2.2, the better performance of the gate rudder was expected for actual sea conditions because of higher loading of the propeller due to waves, wind and hull fouling. According to the sea margin data of the similar vessel (L*B*D*d =110.7*17.4*8.2*5.4), the sea margin of the ship with conventional rudder was expected to be around 25%. The sea margin data for the two vessels were provided by the ship owner[8].

Here, the sea margin (power margin) was calculated by the following simple formulae.

$$SM = (P_{ACT} - P_0) / P_0 \quad \text{-------- (31)}$$

$$PACT = PMES * (DWT/6000)2/3 \quad \text{------- (32)}$$

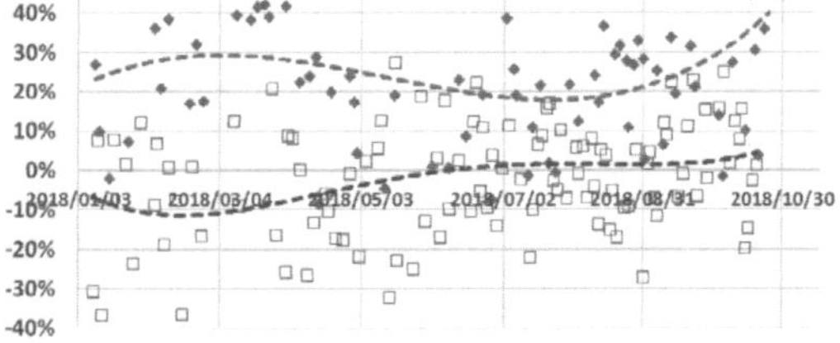

Figure 9 Sea Margin of two Ships
(Sakura: upper, Shigenobu(GR): lower)

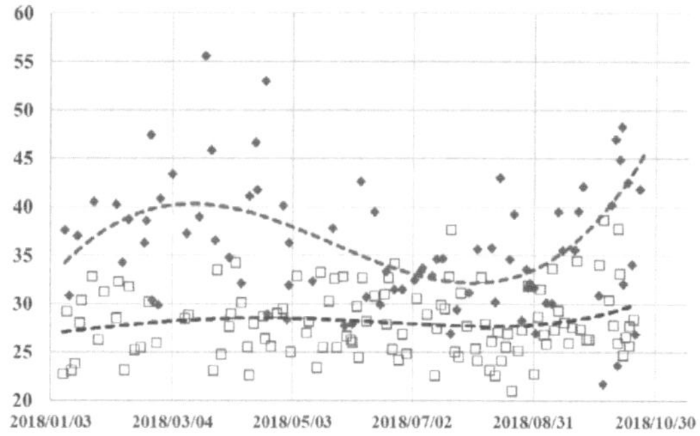

Figure 10 Litre/mile of two Ships
(Sakura: upper, Shigenobu(GR): lower)

Figure 9 and Figure 10 shows the difference of sea margin (power margin) and litre/mile data of the two vessels respectively.

This is a quite an interesting result. On the grounds that the two vessels are running on the same day and the same route (Tokyo-Hachinohe-Tomakomai), it is apparent that the difference of the two vessels has originated from only the rudder systems.

The obtained sea margin from the vessel with the conventional flap rudder is around 26% while the vessel with gate ruder is -1%. The largeat difference between the two vessels occurs during winter, In contrast, in summer when the weather is calm, the difference is very close to the sea trial results of 14%.

And so, the difference of actual sea margin from the calm sea conditions can be regarded as 13% instead of 27%.

Table 4 Summary of Voyage Data

	Vs(kts)	*P(kw)*	*Cadm*	*sea margin*	*litre/mile*
Shigenobu	*13.37*	*2217*	*254*	*-1%*	*28.3*
Sakura	*11.86*	*2044*	*186*	*26%*	*37.6*

Notes: The data does not include a short voyage of less than 10hrs, Sakura(Flap Rudder), Shigenobu(Gate Rudder)

Table 4 shows the averaged voyage data collected from Jan. 2018 to Oct.2018.

5. CONCLUSION

The recent full-scale speed trials with this domestic container vessel have confirmed the superior performance of the Gate Rudder system which has shown 14% reduction during speed trials. In addition to this excellent result, the voyage data revealed that in some situations the gain is much larger at 27% in terms of power margin and 33% for the litre/mile ratio.

The propulsive performance of two rudder systems was theoretically investigated. As a conclusion, the gate rudder can be regarded as a ducted propeller system and the conventional powering procedure, which has been used for the conventional rudder system, will not work properly for this new rudder system especially in the prediction of effective wake. The new powering procedure presented in this paper appears to be working well, however further investigation will be required to establish the accurate powering procedure for the gate rudder system as a new propulsor system.

The authors express our great gratitude for their support from *Nippon Foundation* with this project.

6. REFERENCES

1) Molland AF, Turnock SR., Marine rudders and control surfaces: Oxford: Elsevier

2) Khanfir S, et.al., Manoeuvring characteristics of twin-rudder systems: rudder-hull interaction effect on the manoeuvrability of twin-rudder ships. J Mar Sci Technol. 2011

3) Reichel M., Influence of rudder location on propulsive characteristics of a single screw container ship. Int. SMP'09

4) Stierman EJ., The influence of the rudder on the propulsive performance of ships. Part I. & Part II, Int. Shipbuilding Progress 1989

5) Nakatake K., Ando J., Tamashima M., Tashiro K., Interaction among twin screw propellers , and twin rudders in uniform flow, Proc. of 2nd Int. Sym. on Propeller and Cavitation, 1992

6) Sasaki N., Kuribayashi S., Atlar M., Gate Rudder, INT-NAM 2018

7) Sasaki N., Atlar M., Investigation into propulsive efficiency characteristics of a ship with the Gate Rudder propulsion system, AYO Colloquium Series, 2018

8) Fukazawa M. et.al., Full scale gate rudder performance obtained from voyage data, AYO Colloquium Series, 2018

7. AUTHORS BIOGRAPHY

N. Sasaki holds the current position of visiting professor at University of Strathclyde and involved in several energy saving projects. He was working for Newcastle University, MTI (Subsidiary of NYK), NMRI (National Maritime Research Institute) and SHI (Sumitomo Heavy Industries).

AUTHOR INDEX